D1087732

Catalysis by Electron Donor-Acceptor Complexes

Catalysis by Electron Donor-Acceptor Complexes

—Their General Behavior and Biological Roles—

by

Kenzi TAMARU
Dept. of Chemistry, Faculty of Science
University of Tokyo, Bunkyo-ku, Tokyo 113, Japan

and

Masaru ICHIKAWA
Sagami Chemical Research Center
Sagamihara-shi, Kanagawa-ken 229, Japan

A HALSTED PRESS BOOK

KODANSHA LTD.
Tokyo

JOHN WILEY & SONS
New York — London — Sydney — Toronto

Library of Congress Cataloging in Publication Data

Tamaru, Kenzi, 1923–
Catalysis by electron donor-acceptor complexes, their general behavior and biological roles.
(Kodansha scientific books)
"A Halsted Press book."
1. Catalysis. 2. Acid-base equilibrium.
I. Ichikawa, Masaru, joint author. II. Title.
QD505.T34 1975 541'.395 75-28051
ISBN 0 470-84435-3

KODANSHA EDP NO.: 3043-298702-2253(0)

Published in Japan
by
KODANSHA LTD.
12-21, Otowa 2-chome, Bunkyo-ku, Tokyo 112, Japan
Published
by
HALSTED PRESS
a Division of John Wiley & Sons, Inc.
605 Third Avenue, New York, N.Y. 10016, U.S.A.

PRINTED IN JAPAN

PREFACE

When electron donating and accepting molecules are brought into contact, they generally form electron donor-acceptor (EDA) complexes, and, to various extents, an electron (or electrons) may be transferred from the electron donor to the acceptor. In many cases, the light absorption spectra of these complexes exhibit absorption bands which are uncharacteristic of either of the component molecules. Mulliken and others have successfully developed a quantum mechanical formulation of the theory of special light absorption which demonstrates that the electron donor-acceptor interaction stabilizes the ground state of the complex. Such theoretical formulations have stimulated a great amount of theoretical and experimental research. Most of this work has, however, been concerned primarily with the static properties of the stable complexes, such as spectroscopic studies. For many biological systems, the formation of EDA complexes has been demonstrated and, in other cases, suggested, and much work has been carried out on charge and electron transfer processes in them. Of course, most of these systems are highly complicated; it is, however, generally accepted that such transfer processes probably play an important role in biological systems.

In comparison with the enormous amount of work done on the static or structural properties of EDA complexes, their reactivity or dynamic properties have tended to be neglected somewhat. In many EDA complexes, specific reactivities, similar to the case of light absorption, have been demonstrated. These are markedly uncharacteristic of either of the component molecules, and the formation of electron donor-acceptor complexes is generally associated with the appearance of new catalytic activity. For example, graphite is steel-grey to black in color with a metallic sheen and is very stable and inert, being oxidized only with difficulty even at high temperatures. However, when it is brought into contact with potassium, its color turns to gold or blue, reminiscent rather of the alchemist's dream of converting base metals into noble ones. Actually, such lamellar compounds behave as good catalysts for hydrogenation, hydrogen exchange, the isomerization of hydrocarbons, etc., just as platinum does. The tranquilizer, chlorpromazine, and vitamin K_3, α-methyl naphthoquinone, not only form an EDA complex, but this exhibits a new catalytic activity for hydrogen exchange with C_2D_2. Neither chlorpromazine nor quinone alone is a catalyst for this reaction.

As we can understand from the two examples mentioned above, EDA complexes frequently demonstrate high catalytic activities which cannot be expected of either of the component molecules. Since many electron donors and acceptors are involved in most biological systems, it appears worthwhile to establish a more panoramic view of the whole concept of this new field. A fuller insight into the mechanism of catalytic reactions in general, and particularly those of biological systems, will so be gained. Although we ourselves do not feel competent to review the whole aspect of biological systems in relation to electron donor-acceptor behavior, and have restricted ourselves here to rather typical examples, we believe that further extensions of the concept of electron donor-acceptors will offer a means for gaining new and deeper knowledge of nature.

In this book, an attempt has been made to review all relevant work on EDA complexes and to present it in a coherent manner. The subject matter of the book is surrounded by much controversy, particularly as regards the chemical reactivities and catalytic behavior of EDA complexes and their significance in biological systems.

Following a brief Introduction, Chapter 2 presents a fundamental approach to the chemistry of EDA complexes, with sufficient of the available physicochemical data to illustrate possible applications to catalytic reactions. Chapters 3 and 4 deal with homogeneous and heterogeneous catalysis by EDA complexes, respectively, together with the theoretical background and applications of useful catalysts. Chapter 5 is concerned with the biological and biochemical significance of EDA complexes, attempting to correlate their effects with the physicochemical properties of the complexes, and with their pharmaceutical effectiveness.

The authors are indebted to Mr. W. R. S. Steele and other staff of Kodansha for their editorial and linguistic assistance in the preparation of this book.

Tokyo Kenzi TAMARU
June, 1975 University of Tokyo

CONTENTS

CHAPTER 1

INTRODUCTION

It has been known for a very long time that mixtures of certain electron-donating and -accepting molecules or classes of molecules can form highly colored solids or solutions, a good example being quinhydrone (which actually consists of an equimolar mixture of hydroquinone and benzoquinone) or picrate complexes.

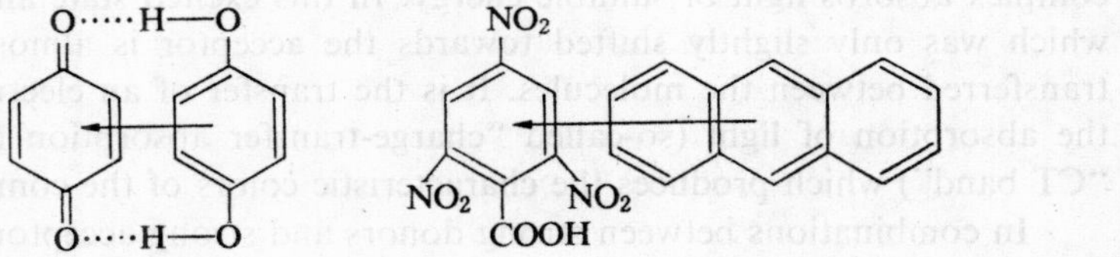

Although the colored material is not a new chemical entity, it possesses characteristic chemical and physical properties different from those of its components. Other changes associated with the formation of these adducts include a marked solubilization of the components in polar solvents, a decrease in diamagnetic susceptibility and an increase in paramagnetic susceptibility of the adducts as compared to the sum of these parameters for the components. Often, spectroscopic properties such as the ultraviolet, visible and infrared spectra of the adducts are slightly or markedly different from the sum of the individual components. Moreover, one finds strikingly simple stoichiometries (usually 1:1 and 1:2) and structures of these adducts.

Such compounds (adducts) have been called molecular complexes or "*Molekülarverbindungen*" since Hantsch and Pfeiffer (1927).[1]

Subsequently, much information on the spectroscopic and thermodynamic behavior of electron donor-acceptor adducts, (molecular complexes) has been collected, and recently comprehensive surveys of such complexes have been published by Briegleb[2] and Andrews and Keefer.[3]

The usual donor-acceptor complexes normally have enthalpies of dissociation of less than 10 kcal/mole, smaller than those of ordinary chemical reactions involving the rearrangement of chemical bonds (typically 50–100 kcal/mole) and similar to hydrogen bonding or van der Waals forces between molecules.

The term "charge-transfer complex" for those molecular complexes was first coined by Mulliken (1952)[4] to describe a certain type of molecular complex with distinctive features. According to Mulliken, the formation of a molecular complex between two molecules, one an electron acceptor and the other an electron donor, relative to each other, causes the above phenomena. Such a complex is stabilized by resonance between no-bond (DA) and dative structures (D^+A^-), the forces involved being called "charge-transfer forces". In the ground state of weak charge-transfer complexes, the two molecules experience the normal physical forces one would expect between two molecules in close proximity, i.e. van der Waals forces, etc., and in addition a small amount of charge is transferred from the donor to the acceptor, which contributes some additional binding energy to the complex. The excited state is formed when the ground-state complex absorbs light of suitable energy. In this excited state an electron which was only slightly shifted towards the acceptor is almost wholly transferred between the molecules. It is the transfer of an electron upon the absorption of light (so-called "charge-transfer absorption band" or "CT band") which produces the characteristic colors of the complexes.

In combinations between strong donors and strong acceptors a comparatively ionic complex may be formed by a different mechanism, such as dipole-dipole and electrostatic interactions between the molecules.

Weiss[5] described the interactions of nitro compounds or quinones (A) with aromatic substances (D·) in terms of a one electron transfer process

$$D\cdot + A \rightarrow D\cdot^+ \cdot A^-$$

leading to the formation of odd electron ions held together by electrostatic forces. Weiss further suggested that the stability of a complex should be dependent on the ionization potential of D: and the electron affinity of A. Certainly the heats of interactions of this kind, usually of the order of a few kilocalories, are much too low to be characteristic of a salt-forming process. Typically, organic molecular complexes are diamagnetic. In the past few years, however, a number of complexes have been found which display paramagnetism and which therefore must have some biradical character.

The term "electron donor-acceptor complex (EDA complex)" has been used frequently in preference to charge-transfer complex by some authors, who reserve the latter term for cases where there is definite evidence of charge-transfer interaction in the ground state. Briegleb and Bent[6] in recent reviews have used the term EDA complex to cover all types of complexes between electron donors and acceptors involving some degree of electron transfer. One criterion given in Bent's review for EDA complexes is that the intermolecular separation of the molecules in the complex is less than one would expect if only dispersion forces were present. We would therefore like to adopt this terminology to include all types of complexes between electron donors and acceptors with various mutual interactions such as charge-transfer, dipole-dipole and electrostatic forces.

The formation of these EDA complexes seems to be a fundamental chemical process, and in fact, in some chemical reactions and catalytic systems, a donor-acceptor complex has been observed and isolated as a reaction intermediate. Thus the properties of EDA complexes may be of wide significance in many fields of chemistry, including catalysis and biochemistry. Nevertheless, there is so far little definite evidence that EDA complexes are of major importance in chemical and catalytic reactions. EDA complexes do, however, possess certain properties which are important in living systems.

The most obvious property is the transfer of charge from one molecule to another. Electron- and charge-transfer or transport systems are vitally important in biological processes. The mechanisms of photosynthesis, oxidative phosphorylation or the various redox reactions are of particular interest in connection with charge transfer. The catalytic mechanisms of these systems are still very unclear, but charge-transfer complexing is a possible method by which molecules involved in such processes could receive and pass on electronic charge.

Another property of electron donor-acceptor forces is that they are relatively long-range as compared to chemical forces. Thus the distance between molecules in these complexes is typically 3.2 to 3.4 Å, whereas chemical bond lengths are less than about 1.5 Å. Charge transfer would therefore facilitate interaction between large, mobile molecules over comparatively long separations.

Charge-transfer forces usually hold the components in the complex together in a specific orientation. This could perhaps act to bring large molecules, including biomolecules, together with their prosthetic groups correctly aligned for interaction.

Many aromatic molecules are good semiconductors, forming donor-acceptor complexes with various partners. For instance, polycyclic aromatic hydrocarbons act as electron donors toward benzoquinones, but as

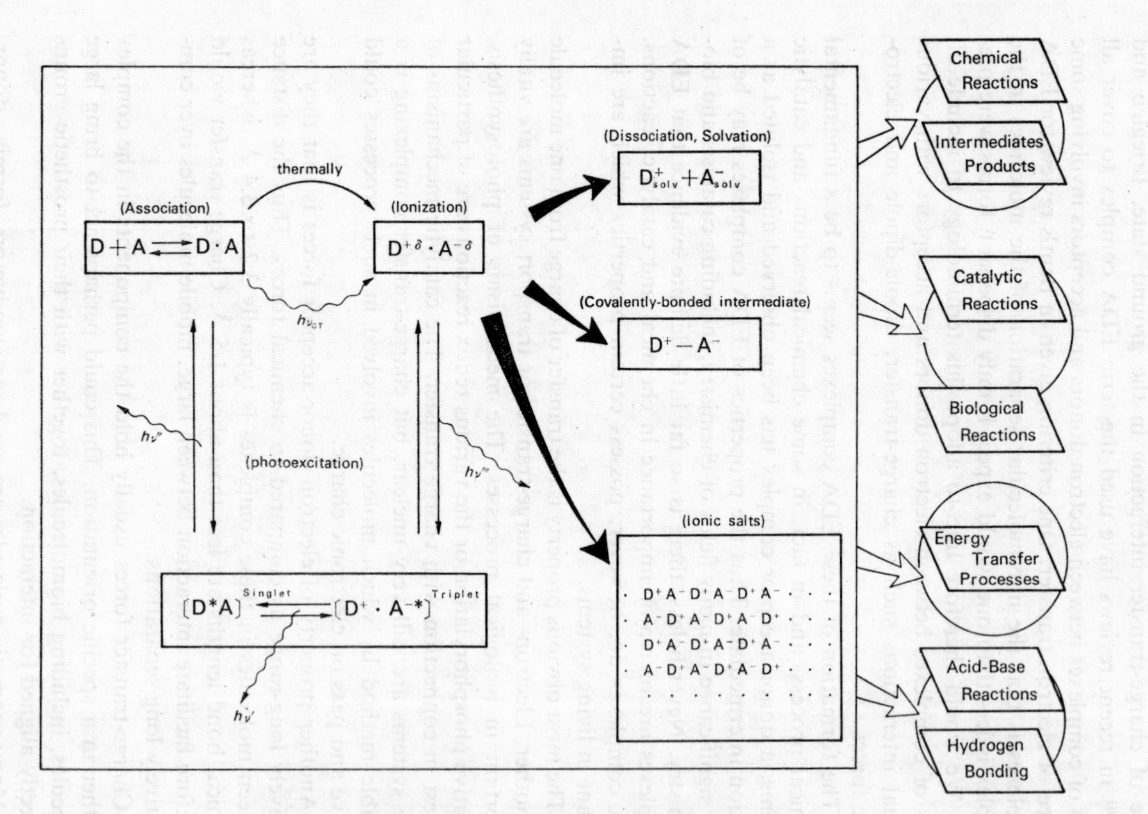

Fig. 1.1 General scheme for the reactions of EDA complexes

electron acceptors toward sodium. One could speculate that the role of electron-transfer complexes may be to control or modify electron density or its flow, like the base of metal oxide semiconductors. The effect of introducing a charge donor or charge acceptor into an electron-conducting system of organic and inorganic solids would be very similar to that of introducing electron-donating or hole-donating impurities into inorganic semiconductors.

The possible role of charge-transfer complexes in reactions has been examined carefully in only a few instances, even though such complexes are frequently postulated as reaction intermediates. The chemical and biochemical significance of EDA complexes has been emphasized by Kosower,[7] Szent-Györgyi[8] and Foster.[9] They classified reactions involving donor-acceptor interactions and complexing into various categories, as shown in Fig. 1.1.

Recently, the laboratories of Inokuchi and Tamaru[10] have carried out extensive work on the homogeneous and heterogeneous catalytic behavior of EDA complexes in connection with the electronic structures of the complexes and their surface properties. Polymerization initiated by charge transfer and electron transport on an electrode surface in the dark, and photo-excitation reactions have also been investigated on the basis of charge-transfer interactions.

We will now briefly consider the part which charge-transfer complexation may play in chemical and catalytic reactions. It may be anticipated that three main categories of reaction processes require consideration: (1) chemical reaction processes (photochemical and thermal), (2) catalytic processes, and (3) energy-transfer processes.

1.1 Chemical Reaction Processes

There are some theoretical studies on the possible role of EDA complexes as intermediates in aromatic electrophilic (or nucleophilic) substitution, rearrangement and exchange reactions. Complexes between the reactants are sometimes identified by spectroscopic methods as reaction intermediates close to the transition states of the reactions. For instance, we shall consider the evidence for rate-determining π-complex formation of aromatic rings with the nitronium ion, i.e. NO^{+}_{2}, in the nitration of aromatic compounds. A charge-transfer complex between iodide and isoprene has been detected by ultraviolet spectroscopy in the halogenation of isoprene. Moreover, a σ complex, i.e. a covalently bonding EDA complex, is observed after short-lived charge-transfer complex formation between the ethoxide ion and 2,4,6-trinitrobenzene in substitution reactions at low temperatures.

A reasonable number of reactions take place in a manner which can often be rationalized in terms of the following steps of complex formation and electron transfer between molecules:

$$D+A \rightarrow (D\ldots A) \rightarrow (D^{+}\text{—}A^{-}) \rightarrow D^{+\cdot}+A^{-\cdot}$$

In the case of electron transfer induced by thermal means (although such reactions may also be light-catalyzed, the overall course is not necessarily the same), such processes can be called thermal electron-transfer reactions.

1.2 Catalytic Processes

The presence of components which could form EDA complexes during catalytic reactions could influence the reactivity and the reaction pathways. Suppose, for example, that the direct reaction between A+B to form P takes place very slowly. If D reacts readily with A to form AD which further reacts rapidly with B to form P and D, as follows, the overall reaction (A+B→P) will proceed readily by repeating the two steps. In this case the donor-acceptor complex (AD) is an intermediate of the reaction process, and the additional component (D) may be regenerated after formation of a product (P).

$$A+D \rightleftarrows AD$$
$$AD+B \rightleftarrows P+D$$

$$A \xrightarrow{+B} P; \quad A \underset{-D}{\overset{+D}{\rightleftarrows}} AD; \quad AD \xrightarrow{+B} P + D$$

The stabilization of the complex (AD) would modify the rate and selectivity of the reaction between the reactant molecules A and B. The net effect might be acceleration or inhibition of the reaction in a particular case.

The stereogeometry of such complexes might affect the reaction pathway to give a specific product different from the product in the absence of complexing reagent. There are many reactions in which a particular pathway is favored because of complex formation between the reactants. Stereospecific hydrolysis by enzymes provides an example, though naturally the forces involved need not be of one particular type. In some instances, however, it seems reasonable or likely that a complex which exhibits charge-transfer light absorption is an actual intermediate in the formation of a certain product. In this class of reaction, the geometry as well as the electronic structure of the donor-acceptor complex are important elements in determining the reaction pathway.

Charge-transfer proceeds between strong donors and acceptors to produce ion-radical pairs or ionic complexes by thermal or photo-excitation. Polar and diradical complexes (in solution as well as in the solid state) may also act as catalytic species, even though the individual components are chemically stable.

1.3 Energy-transfer Processes

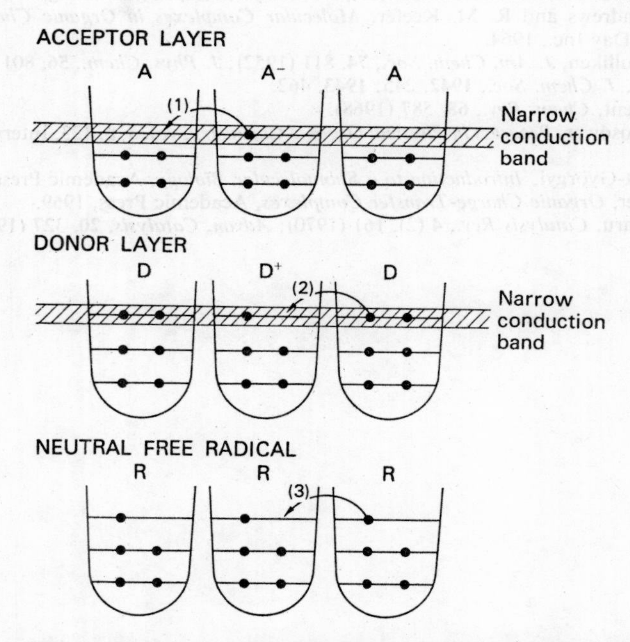

Fig. 1.2 Schematic representation of long-range "electron" or "hole" transfer of donor and acceptor molecules and ions imbedded in a donor layer or an acceptor layer, respectively.

The details of energy-transfer processes are very important to a consideration of the photochemistry of charge-transfer complexes, but have been investigated only to a limited extent.

Generally, charge separation in EDA complexes in the ground state due to thermal and photo-excitation leads to the formation of electrons and holes, which act as electron donors (A^-) and acceptors (D^+), respectively, in relation to other electron donors or acceptors in the molecular crystal layers (. . A.A.A... and . . D.D.D...). Photo-excited donor-acceptor complexes may play a role in the transmission of energy or electron transport by long-range interactions, as indicated schematically in Fig. 1.2, particularly in the redox cycles of some biochemical systems.

REFERENCES

1. P. Pfeiffer, *Die Organische Molekülarverbindungen*, 2nd ed., Ferdinand Enke, 1927.
2. G. Briegleb, *Elektronen Donator-Acceptor Complexe*, Springer-Verlag, 1961.
3. L. J. Andrews and R. M. Keefer, *Molecular Complexes in Organic Chemistry*, Holden-Day Inc., 1964.
4. R. S. Mulliken, *J. Am. Chem. Soc.*, **74**, 811 (1952); *J. Phys. Chem.*, **56**, 801 (1952).
5. J. Weiss, *J. Chem. Soc.*, **1942**, 245; **1943**, 462.
6. H. A. Bent, *Chem. Rev.*, **68**, 587 (1968).
7. E. M. Kosower, *Progress in Physical Organic Chemistry*, vol. 3, p. 123, Interscience, 1956.
8. A. Szent-Györgyi, *Introduction to a Submolecular Biology*, Academic Press, 1960.
9. R. Foster, *Organic Charge-Transfer Complexes*, Academic Press, 1969.
10. K. Tamaru, *Catalysis Rev.*, **4** (2), 161 (1970); *Advan. Catalysis*, **20**, 327 (1969).

CHAPTER 2

Formation of EDA Complexes

2.1 Electron Donors and Electron Acceptors in Charge-Transfer Complex Formation

2.1.1 Classification of electron donors and acceptors

Molecules capable of giving up an electron are defined as electron donors (D), and measurements of their ionization potentials provide a convenient index of donor ability. Usual donor molecules such as aromatic hydrocarbons including alkenes and alkynes, which have π-donating orbitals (π donors), and alkylamines and pyridine, which have non-bonding electrons (n donors), fall in the range $I_p = 5–12$ ev.

Molecules which can accept an electron or electrons are called acceptors (A), and acceptor ability is clearly related to electron affinity and reduction potential. Aromatic nitro compounds and quinones are π acceptors, and halogen molecules having vacant σ antibonding orbitals act as σ acceptors (see Fig. 2.1).

In connection with this classification, it should be kept in mind that in some cases the same molecule can function either as a donor or as an acceptor according to the circumstances. Also, in a large molecule, simultaneous functioning as a donor at one site and as an acceptor at another is possible. Molecules containing π electrons, such as ethylene and benzene, can act as either weak donors or very weak acceptors. Other things being equal, donor ability increases with decreasing ionization potential (I_p) and acceptor ability with increasing electron affinity (EA). Among aromatic hydrocarbons, I_p decreases and EA increases with increasing size of the molecules; graphite, with $I_p = \text{EA}$, is an extreme example and is in fact both a good acceptor and a good donor (Table 2.1)

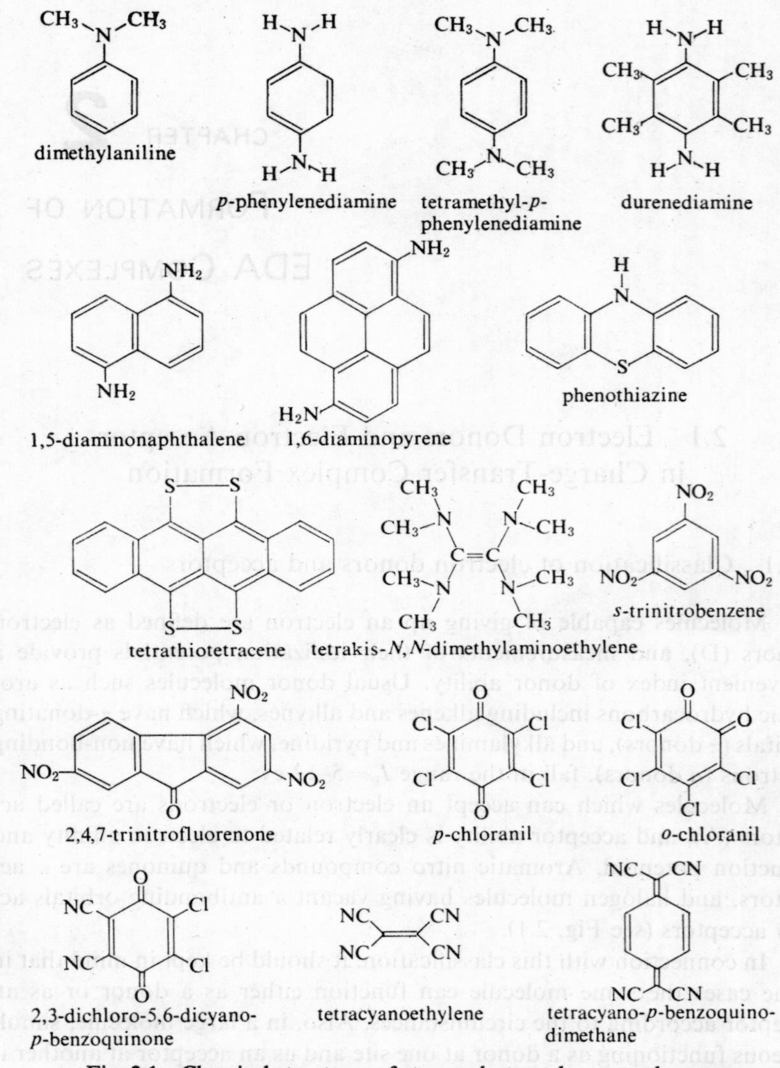

Fig. 2.1 Chemical structures of strong electron donors and acceptors.

With any given unsaturated or aromatic hydrocarbon, its donor or acceptor capability can be strengthened by the introduction of suitable substituent groups. The weak donor property of benzene is strengthened by introducing electron-releasing methyl or amino groups (inductive effect) whereas the four electrophilic cyano groups in tetracyanoethylene (TCNE)

TABLE 2.1 Calculated values of ionization potentials and electron affinities of some polycondensed aromatic hydrocarbons

	Ionization potential (eV)	Electron affinity (eV)
Benzene	9.53	−1.59
Naphthalene	8.52	0.25
Phenanthrene	8.56	0.01
Anthracene	8.16	0.42
Pyrene	8.05	0.42
Tetracene	7.71	0.95

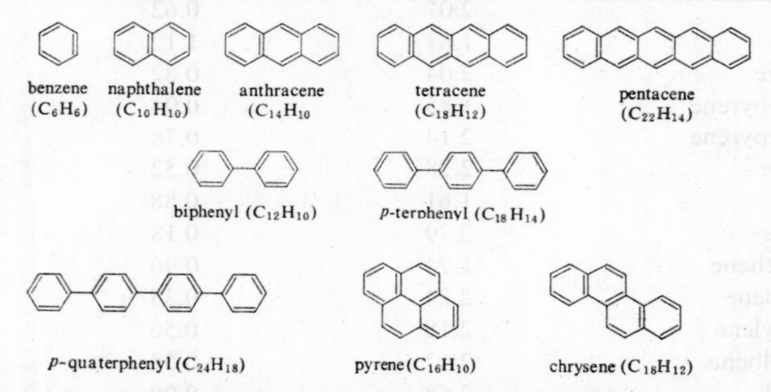

(Source: S. Ehrenson, *J. Phys. Chem.*, **66**, 710 (1962). Reproduced by kind permission of the American Chemical Society, U.S.A.)

greatly increase its acceptor capability compared to ethylene. Table 2.2 presents the electron affinities of various aromatic compounds, as determined experimentally from reduction potentials in polar solvents and photoemission spectra *in vacuo*.

The nitro group is strongly electron-attracting, and as a result both polynitroaliphatic and polynitroaromatic substances have acceptor properties. The polynitroaromatics, as well as other aromatic substances which have strongly electron-withdrawing ring substituents, can be classified as π acids.

A great many colored solid adducts of polynitroaromatic compounds with aromatic donors have been prepared. Some of these molecular compounds are sufficiently stable so that they have characteristic melting points. Picric acid has been used widely in analytical work in the preparation of solid derivatives of aromatic hydrocarbons. Recently 2,4,7-trinitrofluorenone has been used to prepare solid adducts of a wide variety of donors.

Nitrobenzene is, at best, a very weak acceptor. Aluminum chloride, unlike aluminum bromide, apparently does not interact appreciably with aromatic hydrocarbons. Yet nitrobenzene solutions of aluminum chloride, which ordinarily are yellow, change color when hydrocarbons of apprecia-

TABLE 2.2 Electron affinities determined from reduction potentials

Substance	Reduction potentials in DMF[†1] *vs.* s.c.e.[†2] $-E_{\frac{1}{2}}$ (V)	Electron affinity A_G(eV)	Ref.[†3]
Naphthalene	2.50	−0.02	*1, 6*
Anthracene	1.99	0.64	*1, 2, 3*
Naphthacene	1.65	1.08	*4*
Pentacene	1.37	1.45	*4*
Phenanthrene	2.425	0.22	*1, 5, 6*
1,2-Benzanthracene	2.04	0.71	*4, 6*
1,2,3,4-Dibenz-anthracene	2.07	0.75	*1*
1,2,5,6-Dibenz-anthracene	2.06	0.76	*1*
Pyrene	2.07	0.62	*1, 2*
Perylene	1.67	1.12	*1, 2*
Coronene	2.04	0.82	*1*
1,2-Benzpyrene	1.85	0.94	*1, 2*
3,4-Benzpyrene	2.14	0.78	*2*
Chrysene	2.28	0.52	*1*
Azulene	1.61	0.88	*1*
Fluorene	2.79	−0.18	*2*
Fluoranthene	1.74	0.96	*1*
Biphenylene	2.24	0.35	*1*
Triphenylene	2.18	0.56	*1*
trans-Stilbene	2.12	0.55	*3*
Biphenyl	2.58	0.00	*1, 5*
p-Benzoquinone	0.45	1.84	*8*
p-Chloranil	−0.08	2.73	*8*
p-Bromanil	−0.08	3.00	*8*
1,4-Naphthoquinone	0.64	1.85	*8*
9,10-Anthraquinone	0.87	1.78	*8, 10*
9,10-Phenanthraquinone	0.60	2.06	*9*

ble basicity are added. With *m*-xylene, for example, a light orange color is observed, and hexamethylbenzene solutions are deep red. The colors are attributed to 1:1:1 complexes.

Presumably, by coordinating with it, aluminum chloride appreciably enhances the acceptor strength of nitrobenzene. That is, the role of the metallic halide in this type of interaction is regarded as similar to that which it plays in the formation of complexes of the type $ArH \cdot HCl \cdot AlCl_3$.

TABLE 2.2—*Continued*

Substance	Reducation potentials in DMF†[1] *vs.* s.c.e.†[2] $-E_{\frac{1}{2}}$ (V)	Electron affinity A_G(eV)	Ref.†[3]
Tetracyanoethylene	−0.16	2.56	*7, 11*
Tetracyano-quinodimethane	−0.12	2.82	*7, 11*
Benzonitrile	3.25	−0.85	*12*
1,2-Dicyanobenzene	2.63	−0.16	*12*
1,4-Dicyanobenzene	2.48	0.04	*12*
4-Cyanotoluene	3.26	−0.73	*12*
Pyromellitic dianhydride	0.53	1.52	*7*
Phthalic anhydride	1.29	1.11	*7*
Tetrachlorophthalic anhydride	0.84	1.80	*7*
Maleic anhydride	0.90	1.55	*7*
Nitrobenzene	1.10	1.33	*13, 14, 15*
1,3-Dinitrobenzene	0.83	1.64	*13, 14*
1,2-Dinitrobenzene	0.62	1.85	*14*
1,4-Dinitrobenzene	0.75	1.69	*13, 14, 16*
4-Chloronitrobenzene	0.99	1.46	*14, 16*
1,3,5-Trinitrobenzene	0.49	2.07	*13, 14*

†[1] DMF=dimethylformamide.
†[2] s.c.e.=standard calomel electrode $A + e \rightleftharpoons A\cdot^{-} - E_{\frac{1}{2}}$ volts
†[3] References: *1*, Streitwieser and Schwager (1962). *2*, Allison, Peover and Gough (1963). *3*, Wawzonek, Blaha, Berkey and Runner (1955). *4*, Bergman (1954). *5*, Geske and Padmanabhan (1965). *6*, Given (1958). *7*, Peover (1962). *8*, Peover (1962). *9*, Peover (1961). *10*, Jones and Spotswood (1962). *11*, Acker and Hertler (1962). *12*, Rieger, Bernal, Reinmuth and Fraenkel (1963). *13*, Hansen, Toren and Young (1966). *14*, Peover (1964). *15*, Geske and Maki (1960). *16*, Maki and Geske (1961).
(After M. Batley, *Thesis, Univ. of Queensland*, 1966.)

Silver salts and salts of copper(I) and mercury(II), both in the solid state and in aqueous solution, interact with alkenes to form complexes. A large number of solid silver salts, which frequently have sharp melting points, have been prepared in characterizing new types of unsaturated substances. Many solid adducts of unsaturates, notably of dienes, with halides of Pt(II), Pd(II), Rh(II) and related metallic ions have also been prepared, and dienes are known to react with metal carbonyls to form stable solids. In fact, in many instances, the forces of coordination between the components in these solid metal halide complexes of alkenes and dienes approach those established in the formation of strong dative bonds. By investigating the distribution of an alkene between carbon tetrachloride and an aqueous solution as the metallic ion concentration of the aqueous phase is varied, equilibrium constants for formation of a number of 1:1 complexes of Ag^{+}, Cu^{+}, and Hg^{2+}, have been obtained. In summary, the common types of donors and acceptors are listed in Table 2.3.

TABLE 2.3 Common types of donors and acceptors

Donor type	Example	Dative electron†	Acceptor type	Example	Dative electron goes to
n	$:NR_3$	Non-bonding lone pair	v	BCl_3	vacant orbital
π	benzene	Bonding π-orbital	π	TCNE	antibonding σ orbital
			σ	I_2	antibonding σ orbital

† "Dative electron" refers to the electron transferred from donor to acceptor. (Source: R. S. Mulliken and W. B. Person, *Molecular Complexes*, p. 4, 1969. Reproduced by kind permission of J. Wiley and Sons, Inc., U.S.A.)

2.1.2 Formation of EDA complexes

Donors and acceptors may interact, usually weakly, to form complexes, as in a molecular association.

$$D + A \overset{K_c}{\rightleftharpoons} DA$$

In this case the component molecules approach to within about 3.5 Å to associate with each other (e.g. solvation) but little electron transfer occurs between them. There is only weak, if any, covalent bonding between the two components; the forces involved are van der Waals or hydrogen-bonding interactions. In these loose complexes, the identities of the original molecules are to a large extent preserved. The complexes absorb light in a manner different from either the donor or acceptor. For instance, one can observe a new absorption band in a solution of benzene and iodine dissolved in *n*-heptane, a band not appearing in the spectrum of either component alone.

$$DA \xrightarrow{h\nu} D^+A^-$$

The Mulliken theory of charge-transfer complexes suggests that the ground state is stabilized by some contribution (resonance) from the dative state. Stabilization of the ground state of a charge-transfer complex by a contribution from the charge-transferred form implies that, among other things, the constant of complex formation, K_c, and the light absorption process might be related to the donor and acceptor properties of the components of the complexes.

In weak charge-transfer complexes the physicochemical properties of the components do not change much from those of the free states, and only the geometrical arrangements of the donor and acceptor molecules are much affected compared with the free states.

Stronger donors (having lower I_p) and/or stronger acceptors (having larger EA) form stronger charge-transfer complexes, in which the charge-transferred structure is a major contributor to the ground state, i.e. an ionic complex. In an extreme case, complete charge transfer causes charge separation between the donors and acceptors, yielding ion pairs binding by electrostatic and dipole-dipole interactions.

The former type of complex may be called a CT complex (π complex or outer complex) and the latter an ionic complex (σ complex or inner complex) according to the extent of charge transfer between the donor and acceptor components. Among the ionic complexes the structures and electronic properties of the components differ markedly from those of the free molecules, and these complexes act as new chemical species in chemical and catalytic reactions.

Accordingly, we generally designate donor-acceptor adducts as electron donor-acceptor (EDA) complexes, taking the wider definition, including all types of such complexes with various extents of charge-transfer between the donor and acceptor components, as shown below.

$$D + A \overset{Kc}{\rightleftharpoons} (DA) \overset{h\nu}{\rightleftharpoons} (D^+\text{—}A^-) \rightleftharpoons D^+_{solv} + A^-_{solv}$$

Most EDA complexes exhibit resonance among these limited electronic states between non-bonding (DA) and dative structures (D^+—A^-). In a polar solvent ionic complexes may exist as an ion pair of solvated radical anions and cations separated by long distances.

2.2 Preparation of EDA Complexes

Most sudies of EDA complexes have been made in solution and/or in the solid state. Usually solutions of donor and acceptor components are mixed, and a precipitate is easily obtained from concentrated solutions. In order to enhance complex formation it is desirable for both donor and acceptor molecules to approach closely. For instance, although a mixed solution of hydroquinone and parabenzoquinone exhibits a pale yellow color due to formation of a weak charge-transfer complex, the solution becomes deep blue due to the formation of quinhydrone on rapidly freezing the mixed solution, and a characteristic esr signal attributable to the formation of hydroquinone radicals is observed. When the frozen, colored solution is melted, the original pale yellow color is restored and the esr signal disappears. Isenberg and Szent-Györgyi[1)] have carried out similar experiments on the complexes formed between tryptophan or serotonin with FMN (flavin mononucleotide). Freezing these complexes produces a very marked intensification of color which disappears on rewarming. Similar colors can also be produced by preparing solid complexes.

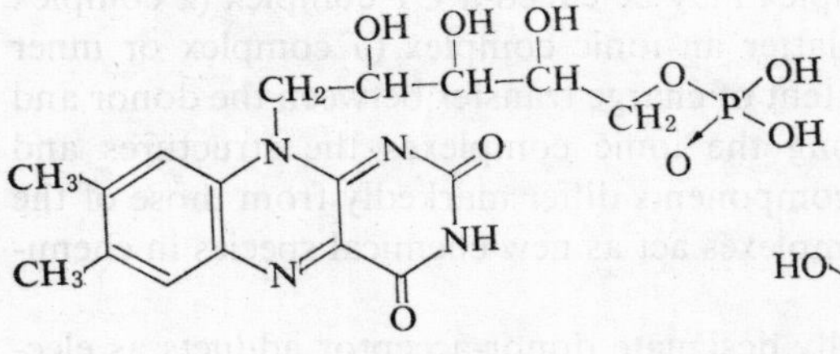

FMN

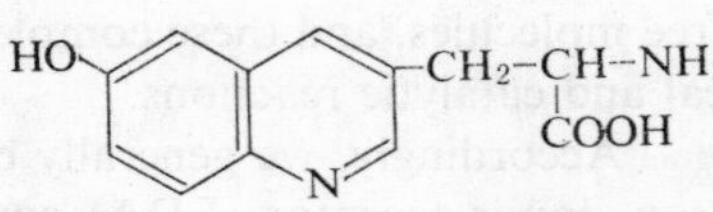

seretonin

On freezing the mixed solution the formation of FMN^- has been detected by esr spectrometry as a result of complete charge transfer from serotonin to FMN in the rigid solvent matrix. Under higher pressures or lower temperatures the spectra of some weaker charge-transfer complexes exhibit shifts to longer wave length, indicating enhancement of charge transfer in the ground state due to closer approach of the components.

It is probably worth pointing out that there are some special features in the interpretation of the spectrum of the blue starch iodine complex, the color being due to interacting triiodide ions rather than iodide alone. From an X-ray analysis of an amylose–I_2 inclusion compound Rundle *et al.*[2)] have reported that the iodine atoms lie linearly at equal distances in the amylose channel, as shown in Fig. 2.2.

The change of the visible spectra of the iodine complex is explicable in terms of charge-transfer phenomena between iodide (I^-) and iodine (I_2) to form triiodide (I_3-) in the amylose matrix. On increasing the length of the amylose chains, the CT band of triiodide shifts to longer wave length (blue shift) in the highly polarized amylose channel.

There are some examples of steric hindrance of charge-transfer interaction due to structural restraints on the close approach of donor and acceptor molecules.[3)]

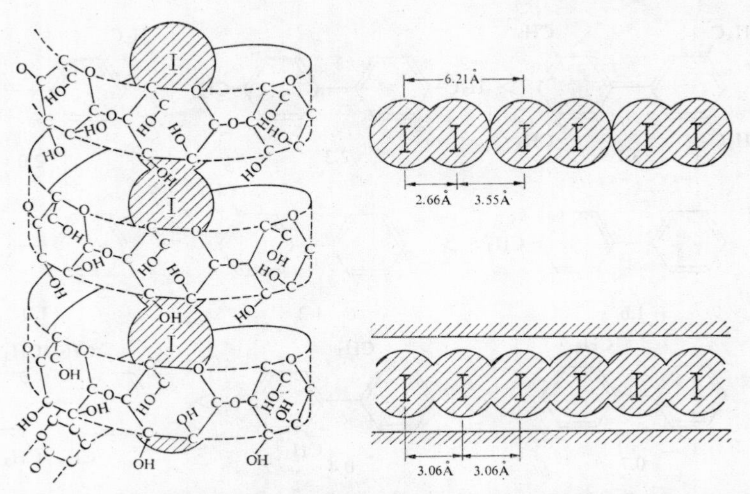

Fig. 2.2 Model of amylose-I_2 inclusion compound based on X-ray analysis. (Source: R. E. Rundle, J. F. Foster and R. Baldwin, *J. Am. Chem. Soc.*, **66**, 2117 (1944). Reproduced by kind permission of the American Chemical Society, U.S.A.)

It has been found that the equilibrium constants, K_c (1/mole),

$$K_c = (\text{ArH} \cdot \text{TNB})/(\text{ArH})(\text{TNB})$$

for formation of 1:1 complexes of *s*-trinitrobenzene (TNB) with a series of biphenyl derivatives (ArH) in carbon tetrachloride at 25°C increase with changes in donor in the order shown. The K_c values are listed below the formulas of the donors. The substitution of methyl groups in the *ortho* positions of biphenyl, as opposed to substitution at *meta* and *para* positions, has an unfavorable influence on the strength of the donor, and the effect is most pronounced when all four *ortho* positions are substituted.

The observed diminution in K_c values with increasing *ortho* methylation most certainly is a reflection of increased hindrance to a coplanar arrangement of the two rings of the donor. In interpreting this observation it has been assumed that the acceptor would interact strongly only with one ring of a biphenyl donor. Apparently the second biphenyl ring, when it is forced to lie far out of the plane of the donor ring, provides interference to nitro groups protruding from the acceptor ring. It is presumed that under these circumstances the acceptor ring is forced out of a favorable parallel orientation with respect to the donor ring or, if a parallel configuration is maintained, that it is forced to lie so far away from the donor ring that effective interaction cannot occur. Similarly, Klemn and Sprague have found that β-substituted cyclohexyl or cyclopentyl compounds of

3.7 > 2.3 > 2.0 >

1.6 > 1.2 > 1.0 >

0.7 > 0.4 > <0.1 ~

<0.1

naphthalene form more stable 1:1 CT complexes with 2,4,7-trinitrofluorenone than do α-substituted derivatives.[4] This might be because the substituent groups and the peri-hydrogen of naphthalene, as shown below, are too close to permit the formation of a planar compound, inhibiting overlapping of the naphthalene and fluorenone moieties due to steric hindrance. In complexes between weak donors and weak acceptors the stabilization and the geometry of the complexes greatly depend upon steric factors rather than the electronic properties. Methylated benzenes such as xylene and hexamethylbenzene produce relatively weak complexes with Ag^+,[5] and *cis*-2-butene forms a stronger CT complex than *trans*-2-butene with Ag^+ and $AlCl_3$.[6]

α-cyclohexyl (or -pentyl) naphthalene

There are examples of the optical resolution of racemic mixtures based on the different stabilities of the charge-transfer complexes formed with optically active electron donors or acceptors. The acceptor (−)-α-(2,4,5,7-tetranitro-9-fluorenylideneaminooxy)-propionic acid[7] (II) has been employed in the separation of optical isomers of α-naphthyl-*sec*-butyl ether, certain asymmetric phosphoryl compounds,[8] and hexahelicene[9] (I). The last is one of an interesting group of compounds which are asymmetric because of intramolecular overcrowding and consequent non-coplanarity of the rings. A partial resolution of α-naphthyl-*sec*-butyl ether has also been accomplished by chromatography using (I) impregnated on silicic acid as the adsorbent.[10]

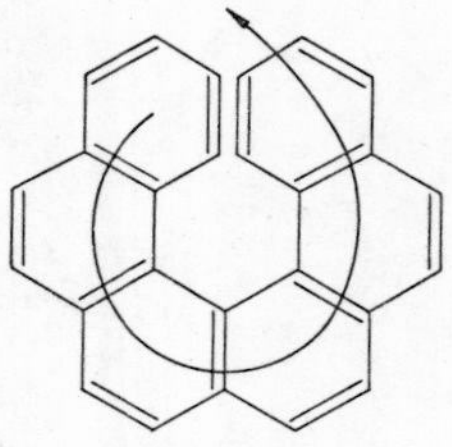

hexahelicene

(I)

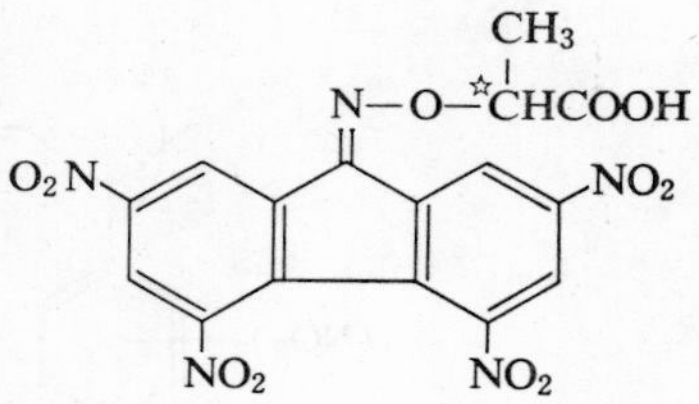

CH_3 N—O—CHCOOH O_2N NO_2 NO_2 NO_2

d,l-sec-butyl-2,4,7-trinitrofluorenyl derivative

(II)

The selective absorption of different donor species on chromatographic stationary phases containing electron acceptors has been used to estimate the degree of association of the acceptor with a given donor; the acceptors include iodine,[11] trinitrofuluorenone (TNF),[12] and 1,3,5-trinitrobenzene.[13] The same principle may be applied to separate mixtures of donors. The converse procedure of separating electron acceptors on a stationary electron-donor phase has been used.[14a]

Norman[14b] has described the use of TNF, in the form of a liquid coated on firebrick, as a stationary phase up to temperatures of 200°C in gas-liquid chromatography. Thin-layer chromatography (TLC), using silica gel impregnated with TNF, has been employed to separate hydroaromatic compounds derived from phenanthrene, anthracene and chrysene.[15] Other separations of aromatic hydrocarbons have been made using thin layers of silica gel containing 1,3,5-trinitrobenzene,[14] and picric acid.[16] The behavior of aromatic amines on thin layers containing 1,3,5-trinitrobenzene or picryl chloride has also been investigated.[17]

Column chromatography was used by Godlewicz[18] to separate hydrocarbon mixtures. Silica gel was impregnated with 1,3,5-trinitrobenzene, with the intention that the latter compound should act as a color-indicator for the various hydrocarbons. It is probable, certainly in the light of more recent observations, that the 1,3,5-trinitrobenzene in fact plays an important role in the actual separation of the hydrocarbon mixture, as suggested by Harvey and Halonen.[15] Columns of silicic acid containing picric acid or TNF,[19a] and columns of tetrachlorophthalic anhydride and of tetrachlorophthalimide[19b] have been used. It is of interest that, whereas silica gel alone is relatively ineffective in separating mixtures of closely related aromatic hydrocarbons, silica gel containing 5% TNF provides a clean separation.[15] Amongst the homogeneous materials which have been used as the stationary phase in column chromatography is the polynitrostyrene referred to as nitrobenzylpolystyrene (III).[20]

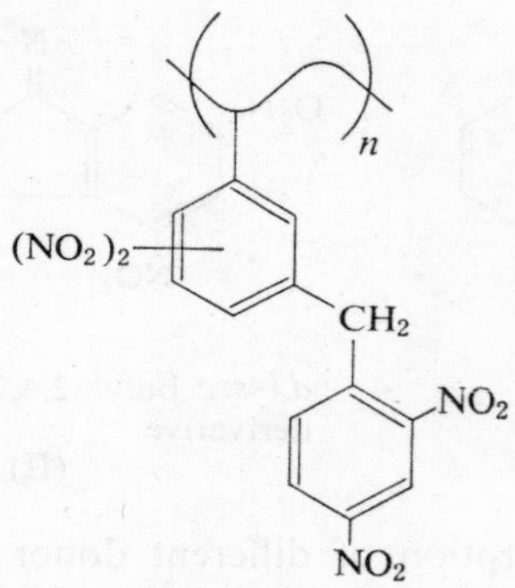

$(NO_2)_2$ — CH_2 — NO_2 — NO_2 — n

(III)

On the other hand, in order to enhance the extent of charge transfer, one can try to synthesize intramolecular donor and acceptor moieties linked with carbon chains. For example, a series of model compounds in which aniline was directly linked to nitrobenzene was synthesized and the spectral properties studied by White.[21]

White has presented evidence that the capacity of a π donor–nitroaromatic acceptor system to display charge-transfer absorption is not strongly influenced by changes in the orientation of the components with respect to each other. The electronic spectra of a series of compounds which have both *p*-aminophenyl and *p*-nitrophenyl groups (a–e) in 1:1 methanol–water solutions have been compared with those of *p*-toluidine and *p*-nitrotoluene.

$H_2N-C_6H_4-(CH_2)_X-C_6H_4-NO_2$ (IV)

H_2N NO_2

(V)

$x=1, 2, 3$ in a, b, and c, respectively

d is the *cis* isomer
e is the *trans* isomer

In all cases compounds a–e showed a strong ultraviolet absorption peak not characteristic of the separate chromophores (Table 2.4). The spectral data which are reported are corrected to eliminate the absorption characteristic of these separate chromophores. The orientational possibilities for the donor and acceptor rings vary widely over this series of multifunctional

TABLE 2.4 Ultraviolet absorption maxima of nitroaminoaromatic compounds

Compound	λ_{max} (nm)	ϵ_{max}	Relative integrated intensity (280–650 nm)
a	324	1620	115
b	313	1330	114
c	310	1480	100
d	312	2420	155
e	308	2470	168

compounds. In one case (compound d) the rings are approximately face to face. In the *trans* isomer of d (compound e) the opportunity for inter-ring overlap is restricted to the 2 positions. The wave lengths of the charge-transfer absorption peaks are relatively invariant over this series of compounds, but the extinction coefficients vary widely.

Recently de Boer *et al.*[22] have comprehensively studied the charge-transfer stabilization of various types of intramolecular model compounds. They measured the visible spectra and phosphorescence of intramolecular CT complexes synthesized from donor and acceptor moieties linked by methylene chains

$$D-(CH_2)_n-A \quad (n=1, 2\cdots\cdots 5)$$

D ≡ , , $-NMe_2$, etc.

$A \equiv$ -N (O, O) –X, -N (O, O, NO_2, NO_2), etc.

For example, the visible spectra of model compounds of dimethylaniline (D) and dinitronaphthalene-*N*-phthalimide (A) linked by methylene chains $-(CH_2)_n-$ ($n=1$, 2, 3, and 4) show a shift of the peaks from 450 nm to 500–550 nm (charge-transfer bands) with increasing carbon number of the methylene chain ($n=2$, 3, 4, 5), and the absorption coefficients of the CT band (at 500–550 nm) are greatest for the model compound with $n=3$, as shown in Fig. 2.3. With longer methylene chains such as $n=4$, the intramolecular charge-transfer interaction decreases considerably, probably due to the relatively unrestricted conformation of carbon chains binding the donor and acceptor moieties. The methylene chain with $n=3$ permits the maximum overlap of the donor and acceptor molecular planes.

Similar experiments have been carried out with model compounds of pyrene (D) and dinitronaphthalene-*N*-phthalimide (A), where the com-

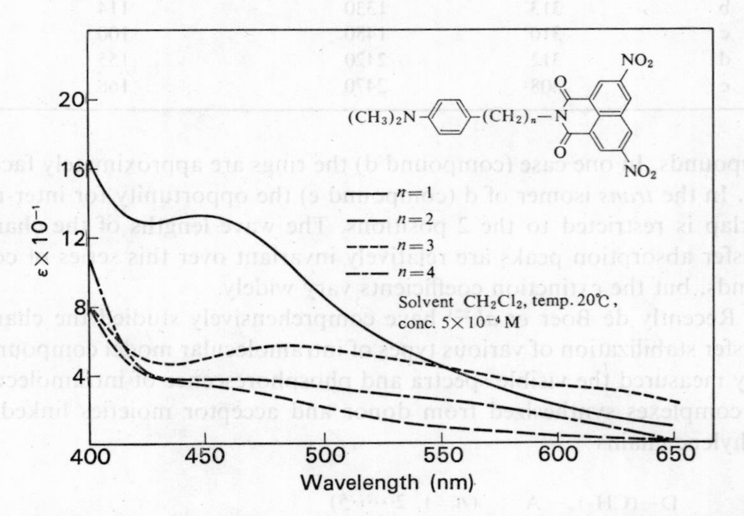

Fig. 2.3 Visible spectra of dimethylanilinyl-*N*-dinitronaphthylimide model compounds with methylene chains ($n=1$, 2, 3, and 4) in methanol.
(Source: H. A. H. Craenen, J. W. Verhoeven and J. de Boer, *Rec. Trav. Chim. Pays-Bas*, **91**, 405 (1972). Reproduced by kind permission of the Royal Netherlands Chemical Society, Netherlands.)

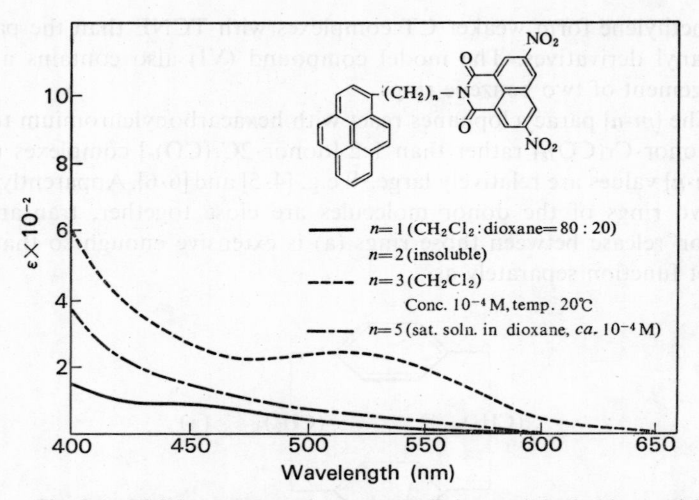

Fig. 2.4 Visible spectra of pyrene-*N*-dinitronaphthylimide model compounds with methylene chains (n=1, 2, 3, and 4) in methanol. (Source: H. A. H. Craenen, J. W. Verhoeven and J. de Boer, *Rec. Trav. Chim. Pays-Bas*, **91**, 405 (1972). Reproduced by kind permission of the Royal Netherlands Chemical Society, Netherlands.)

pound with $n=3$ also exhibited the greatest CT interaction, as indicated in Fig. 2.4.

Cram *et al.*[23] have synthesized fairly stable 1:1 CT complexes between paracyclophanyl derivatives and TCNE. The measurement of the complex stabilities as a function of chain length of the methylene in the cyclophanyl derivatives leads to the conclusion that the CT interaction becomes maximum by overlapping the two benzene rings at close distances, probably due to greater delocalization of electron density of the paracyclophanes in the complexes. Open-chain model compounds such as di-*p*-propylbenzene-

CN CN
C=C
CN CN

$(CH_2)_n$ $(CH_2)_m$

$n, m = 1, 2 \cdots\cdots 6$

(VI)

C_3H_7

$(CH_2)_6$ (TCNE)

C_3H_7

(VII)

hexamethylene form weaker CT complexes with TCNE than the paracyclophanyl derivatives. The model compound (VI) also contains a tight arrangement of two benzene rings.

The [*m*·*n*] paracyclophanes react with hexacarbonylchromium to give 1:1 [donor·$Cr(CO)_3$] rather than 1:2 [donor·$2Cr(CO)_3$] complexes unless the [*m*·*n*] values are relatively large,[24] e.g. [4·5] and [6·6]. Apparently when the two rings of the donor molecules are close together, transannular electron release between those rings (a) is extensive enough so that they cannot function separately as

$(CH_2)_m$ $(CH_2)_n$ (a)

Cr

CO CO CO

$(CH_2)_m$ Cr $(CH_2)_n$ (b)

coordination sites. It is interesting that efforts to prepare compounds of the type [*m*·*n*] paracyclophanechromium (b) by expulsion of carbon monoxide from the monochromium complexes (a) have been unsuccessful even in the cases of the [4·4] and [6·6] donors in which the ring-ring separation distances are favorable. The transformation of a complex of structure (a) to one of structure (b) requires rotation of the complexed ring through 180° about the line joining the 1 and 4 positions. In normal situations rotation of this kind is possible in [4·4] and [6·6] paracyclophanes.

In contrast to complexes with neutral ground states, EDA complexes with dative structures show charge-transfer bands which are markedly sensitive to the nature of the solvent. In fact, the electronic transition in charge-transfer complexing may be the most solvent-sensitive example previously known.

In polar solvents the dative structures of the complexes are generally enhanced due to orientation of the solvent molecules stabilizing the ionic state of the components of the complex, as in the following scheme.

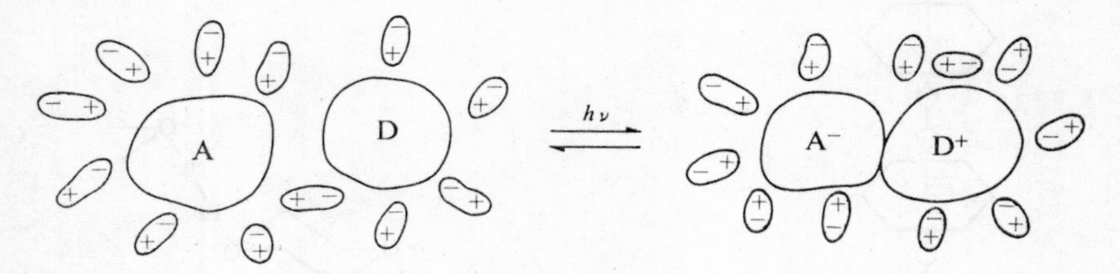

The stabilization energy due to the solvation of D^+ and A^- in polar solvents can be calculated approximately using Born's equation, where ε, R_{A^-} and R_{D^+} denote the dielectric constant of the solvent, and the ionic radii of the A^- and D^+ ions, respectively.

$$-\Delta H_{\mathrm{solv}} = \frac{e^2}{2}\left(1-\frac{1}{\varepsilon}\right)\left(\frac{1}{R_{A^-}}+\frac{1}{R_{D^+}}\right)$$

Taking $\varepsilon = 80$ in aqueous solution and R_{A^-} and R_{D^+} as 5 Å, the stabilization energy can be calculated to be about 55 kcal/mole per ion. The total solvation energy of the ion pair amounts to about 110 kcal/mole in water. Thus, polar solvents strongly stabilize such complexes. For example, TMPD (tetramethyl-*p*-phenylenediamine) forms a greenish CT complex with chloranil in benzene or *n*-hexane, but easily yields an ion-pair complex (TMPD cation and chloranil anion) which exhibits strong esr signals in ethanol, a polar solvent.

2.3 Structures and Physical Properties of EDA Complexes

2.3.1 Crystal structures of EDA complexes

Studies on the crystal structures of EDA complexes have been carried out mainly by X-ray analysis; details should be sought in the references cited below.

CT complexes of benzene with bromine and chlorine molecules form a linear structure,[25] as shown in Fig. 2.5, where the distance between benzene and halogen is constant and smaller than the sum of the van der Waals radii of the components. For the benzene–bromine complex, the

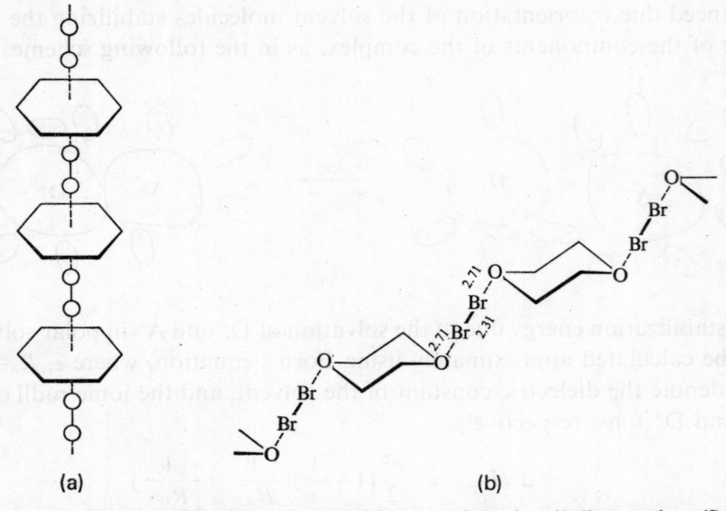

Fig. 2.5 (a) Crystal structure of benzene–bromine (1:1) complex. (Schematic drawing of the perpedicular separation of the molecules in one stack through the unit cell.)
(b) Bridged structure of the dioxane–bromine (1:1) CT complex.
(Source: O. Hassel and J. Hvoslef, *Acta. Chem. Scand.*, **8**, 873 (1954). Reproduced by kind permission of Acta Chemica Scandinavica, Sweden.)

bromine bridge is located on the C_6 axis of the benzene ring, in apparent agreement with the results of Ferguson[26)] based on the infrared spectrum of the benzene–iodine complex.

Groth and Hassel[27)] have suggested that the bromine bridge in the 2:1 methanol–bromine complex (m.p. $-66°C$) is comparable in bond length to a hydrogen bond, since it occupies an equivalent position in the lattice of the complex. The bromine–bromine distance of 2.29 Å is only slightly larger than that in the bromine molecule itself (2.28Å), implying that the extent of charge transfer in the ground state is small. In certain complexes of good lone-pair donors with halogen molecules, appreciable lengthening of the halogen molecular bond is found. Table 2.5 gives X-ray data for

TABLE 2.5 X-ray data for benzene–halogen (1:1) complexes

	Benzene–halogen distance (Å)	Sum of the van der Waals distance (Å)	X–X (Å) in complex	X–X (Å) free molecule
C_6H_6–Cl_2	3.28	3.50	1.99	1.99
C_6H_6–Br_2	3.36	3.65	2.28	2.28

TABLE 2.6 Halogen–halogen distance in EDA complexes with *n*-donors†[1]

Complex	δ†[2]	$(D\text{–}X)_{obs}$	D–X (sum of covalent radii)	D–X (sum of vdW radii)†[3]	$X\text{–}Y_{obs}$	$X\text{–}Y_{free}$
Diselenane–I_2	0.33	2.83	2.50	4.15	2.87	2.67
Dithiane–I_2	0.50	2.87	2.37	4.00	2.79	2.67
Dioxane–2ICl	0.58	2.57	1.99	3.55	2.33	2.32
Dioxane–Br_2	0.91	2.71	1.80	3.35	2.31	2.28
Dioxane–Cl_2	1.02	2.67	1.65	3.20	2.02	1.99
$(CH_3)_3N$–I_2	0.24	2.27	2.03	3.65	2.84	2.67
$(CH_3)_3N$–ICl	0.27	2.30	2.03	3.65	2.52	2.32
Py–ICl†[4]	0.23	2.26	2.03	3.65	2.51	2.32
γ-Pic–I_2†[5]	0.28	2.31	2.03	3.65	2.83	2.67
HMT–$2Br_2$†[6]	0.32	2.16	1.84	3.45	2.43	2.28

†[1] All distances are in Å.
†[2] The term δ means $(D\text{–}X)_{obs}$ − (sum of D and X covalent radii).
†[3] vdW radii = van der Waals radii.
†[4] Py = pyridine. †[5] γ-Pic = γ-picoline. †[6] HMT = hexamethylenetetramine.

benzene–halogen (1:1) complexes. Strong *n* donors such as alkyl amines and pyridine react with halogens to give very stable 1:1 complexes. The halogen–halogen distances in these complexes are markedly lengthened, as shown in Table 2.6, probably due to considerable charge-transfer from non-bonding electrons of *n* donors to the vacant antibonding orbitals of halogens. The crystal structure of pyridine–ICl complex has been determined by X-ray analysis[28] and the N···I–Cl bond is linear, with an N–I distance of 2.26 Å. This configuration of the complex in the crystal state can be explained from theoretical considerations on the maximum orbital overlap between the *n* donor and σ acceptor.

On the other hand, for the benzene–$AgClO_4$ complex it has been found that the double-bond distance in benzene decreases from 1.39 Å in the free molecule to 1.35 Å, whereas the double bonds at the sides increase to 1.43 Å, implying that the benzene ring becomes slightly distorted (Fig. 2.6 (a)).[29] An examination of the crystal structure of the anthracene–1,3,5-trinitro-benzene complex revealed no unusual dimensions of the components (Fig. 2.6 (b)).[30]

The crystal structures of two different complexes involving phenol-

(a)

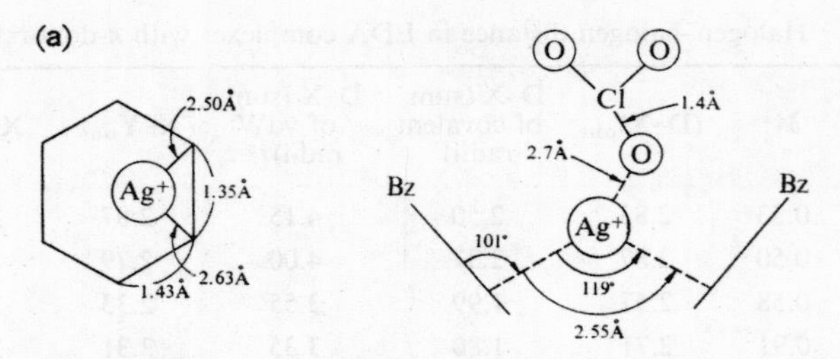

Fig. 2.6 (a) Schematic drawing in projection of the structure of the $Ag^+Bz \cdot ClO_4^-$ crystal.
(Source: R. S. Mulliken and W. B. Person, *Molecular Complexes*, 1969. Reproduced by kind permission of J. Wiley and Sons, Inc., U.S.A.)

(b)

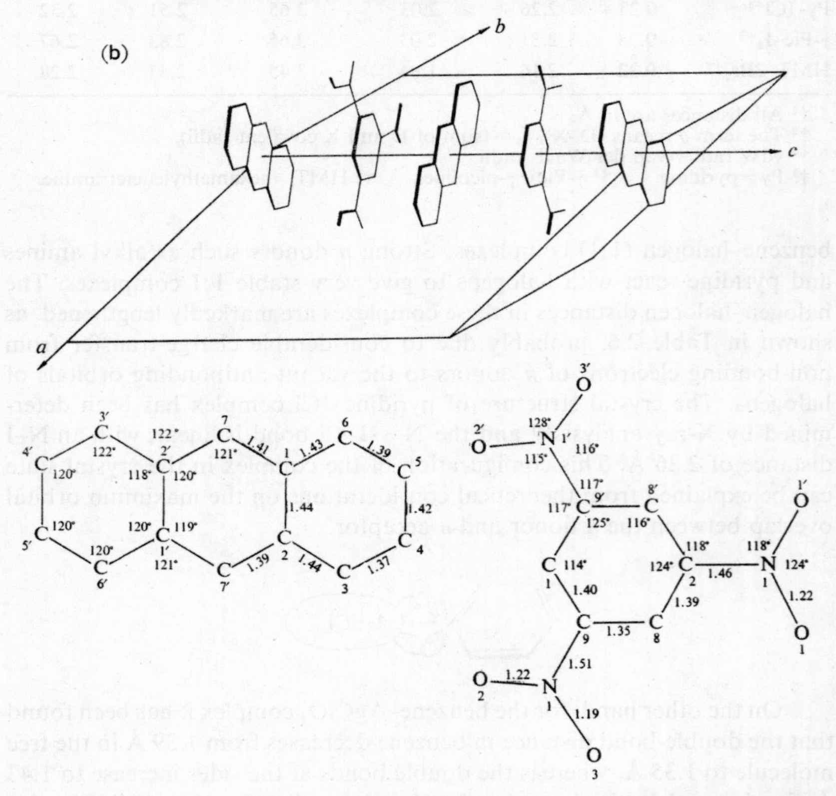

Fig. 2.6 (b) Part of the cell constants of the anthracene-S-trinitrobenzene complex, showing the stacking of molecules along the *c* axis and the molecular dimensions of the components (unit, Å).
(Source: D. S. Brown, S. C. Wallwork and A. Wilson, *Acta Cryst.*, 17, 168 (1964). Reproduced by kind permission of the International Union of Crystallography, England.)

like molecules have been determined by X-ray crystallography.[31] The riboflavin hydrogen bromide–hydroquinone complex (1:1) exists in two allotropic forms (Fig. 2.7 (a, b)). One allotropic form has the hydroquinone overlapping the 8, 9 and 10 positions of the isoalloxazine ring with a separation of 3.35 Å. In the other allotropic form there is a separation between the phenol and the flavin of 3.28 Å and the hydroquinone overlaps the 2 and 3 positions of the isoalloxazine ring. In both forms there is overlap with electron-deficient regions. The electron densities of these mole-

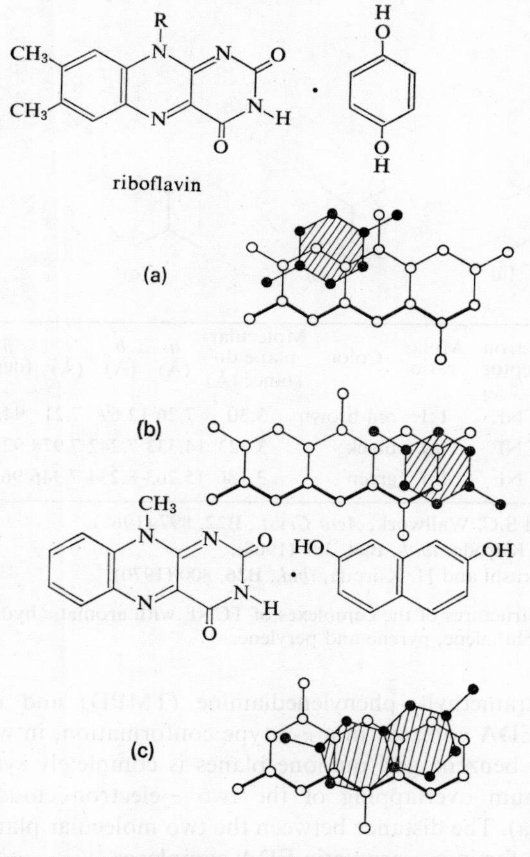

Fig. 2.7 **(a,b) Structure of the allotropic forms of the complex between riboflavin dihydrogenbromide and hydroquinone.**
(c) Structure of the complex between 10-methyl isoalloxazine bromide and sesquinaphthalene-2,7-diol monohydrate.
(Source: C. A. Langhott and C. J. Frichie, *Chem. Commun.*, 1970, 21. Reproduced by kind permission of the Chemical Society, England.)

cules have been calculated by B. Pullman and A. Pullman.[32)] The structure of the molecular complex formed between 10-methyl isoalloxazine bromide and sesquinaphthalene-2,7-diol monohydrate (Fig. 2.7 (c)) is rather different from that of the hydroquinone complex. Overlap occurs mainly in the N-5, C-8, N-10, C-6 region of the isoalloxazine ring, a region of low electron density.

Complexes between donors and acceptors should come close to achieving maximum overlap of the electron clouds of the donor and acceptor molecules. In fact, the tetracyanoethylene lies on the planes of π-donor molecules such as naphthalene, pyrene and perylene,[33)] as shown in Fig. 2.8.

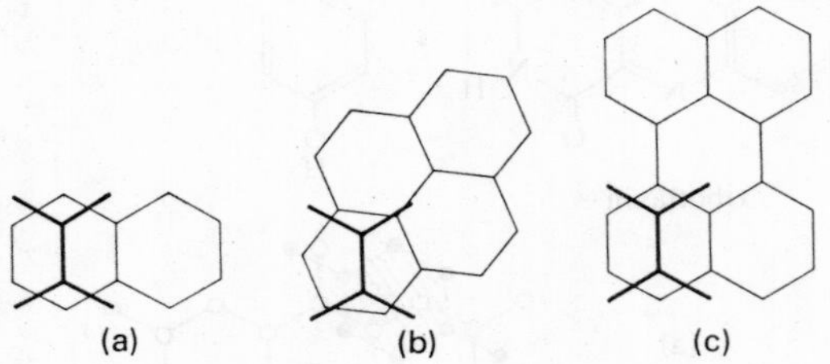

(a) (b) (c)

Electron donor	Electron acceptor	Molar ratio	Color	Molecular-plane distance (Å)	a (Å)	b (Å)	c (Å)	β (deg)	Space group
(a) Naphthalene†[1]	TCNE	1:1	red-brown	3.30	7.26	12.69	7.21	94.4	$C2/m$
(b) Pyrene†[2]	TCNE	1:1	black	3.323	14.333	7.242	7.978	92.36	$P21/a$
(c) Perylene†[3]	TCNE	1:1	green	3.186	15.763	8.234	7.346	96.4	$P21/a$

†[1] R.M. Williams and S.C. Wallwork, *Acta Cryst.*, **B22**, 897 (1967).
†[2] I. Ikemoto and H. Kuroda, *ibid.*, **B24**, 383 (1968).
†[3] I. Ikemoto, Y. Yakushi and H. Kuroda, *ibid.*, **B26**, 800 (1970).

Fig. 2.8 Structures of the complexes of TCNE with aromatic hydrocarbons such as naphthalene, pyrene and perylene.

N,N,N′,N′-tetramethyl-*p*-phenylenediamine (TMPD) and chloranil form a 1:1 ionic EDA complex in a π-π type conformation, in which the orientation of the benzene and quinone planes is completely symmetric, permitting maximum overlapping of the two π-electron clouds,[34a)] as shown in Fig. 2.9(a). The distance between the two molecular planes is the shortest known so far in π-π aromatic EDA complexes.

On the other hand, in crystals such as hexamethylbenzene–chloranil, the longest wave length absorption is observed for light with the electric vector perpendicular to the benzene ring, and this absorption probably corresponds to the charge-transfer band found in solution. It is thus un-

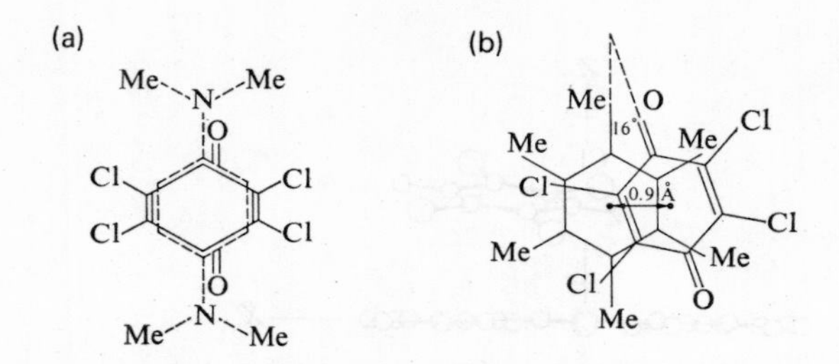

Fig. 2.9 Structures of (a) the strong complex between TMPD and chloranil and (b) the weak CT complex between hexamethylbenzene and chloranil. (Source: C. A. Bear, J. M. Waters and T. N. Waters, *Chem. Commun.*, **1970**, 702. Reproduced by kind permission of the Chemical Society, England.)

likely that the crystalline complex differs greatly from the complex in solution with respect to molecular dimensions. In the hexamethylbenzene–chloranil complex, in which intercomponent hydrogen bonding does not occur, the donor and acceptor rings are apparently considerably less tilted out of a perpendicular orientation with respect to the line joining their centers than are the components of phenoquinone,[34b)] but the alternating rings differ in orientation by about 16° as shown in Fig. 2.9(b). The fact that the rings are not directly superimposed is attributed to the bulkiness of the methyl and chlorine substituents of the donor and the acceptor.

In the chloranil complex of *N*,*N*,*N*′,*N*′-tetramethyl-*p*-phenylene-diamine, the center of the ring of one component lies over that of the other, and the carbonyl oxygens of the chloranil molecule lie directly over the nitrogen atoms of the *p*-dimethylamino substituents of the adjacent donor molecules. The conditions for donor-acceptor interaction are "ideal" in this case, and the inter-ring separation (3.26 Å) is unusually low for a chloranil complex. The ring separation distance in the bromanil complex of the diamine is also low (3.31 Å). It is interesting that these are complexes which display paramagnetism. Apparently these two complexes are significantly ionic in character (D^+A^-), a fact which accounts for the relatively short distance between their donor (D) and acceptor (A) rings.

The structures of charge-transfer complexes of π acceptors with metal 8-hydroxyquinolates are of particular interest. The 1:1 complex between bis(8-hydroxyquinolinato)–Pd(II) and chloranil shows a structure of approximately parallel donor and acceptor molecules (Fig. 2.10(a)). It has been suggested[35a)] that specific interaction between the chlorine atoms and palladium atoms affects the relative orientation of the two components, which is not what would be expected from the overlap and orientation principle. The crystal structure of a complex between 7,7,8,8-tetracyano-quinodimethane and bis(8-hydroxyquinolato)–Cu (II) has been the subject

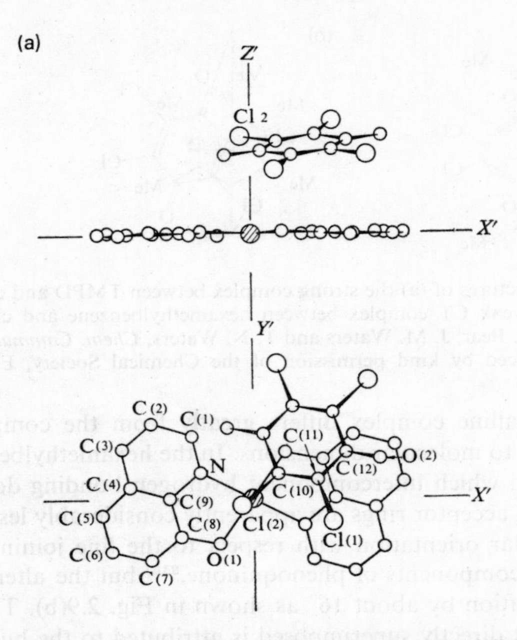

Fig. 2.10 (a) A chloranil molecule in bis(8-hydroxyquinolinato)–Pd-(II)–chloranil complex projected parallel to, and perpendicular to the least-squares best plane relative to the donor molecule.
(Source: B. Kamenar, C. K. Prout and J. D. Wright, *J. Chem. Soc.* (A), 1967, 469. Reproduced by kind permission of the Chemical Society, England.)

of a very careful study by Williams and Wallwork.[35b)] The general features of the structure are indicated in Fig. 2.10(b). The molecules are in the usual plane-to-plane arrangement of alternate components. These are orientated in such a manner that the double bond adjacent to one dicyanomethylene group of the acceptor lies over the 5:8 position of a donor molecule, whilst the other double bond of the acceptor is similarly placed with respect to the benzenoid ring of the centrosymmetrically related donor molecule.

Although it is not known how large the effect of charge transfer is upon bond distance, it is expected that some change in bond distances should result from appreciable charge transfer. In an extreme case, for example, the 1:1 ionic complex of triethylamine and iodine reforms a chemical bond to yield the triethyliodoammonium cation and iodide anion as an ionic salt.[36)]

$$\mathrm{Et_3N} + \mathrm{I_2} \rightleftharpoons \left[\mathrm{Et_3N} \cdots \mathrm{I{-}I}\right] \longrightarrow \mathrm{Et_3N^{+}{-}I} + \mathrm{I^{-}}$$

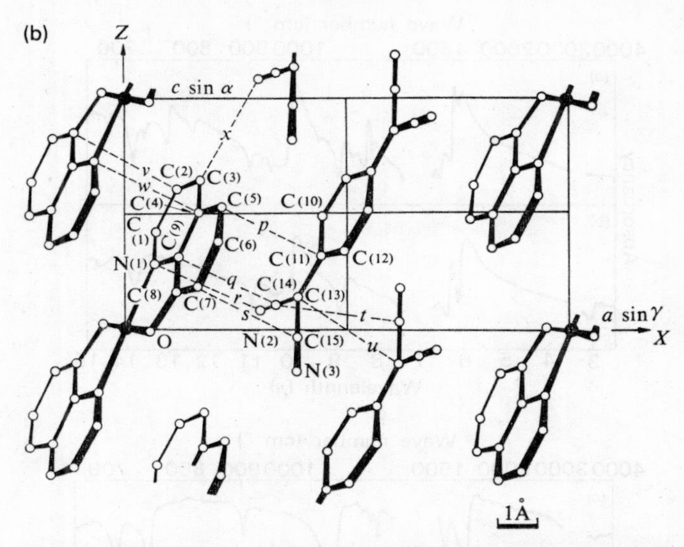

Fig. 2.10 (b) The structure of the bis(8-hydroxyquinolinato)–Cu(II)–7,7,8,8-tetracyanoquinodimethane complex projected along the *b* axis, showing short intermolecular constants. The distances (Å) indicated are $p=3.236$, $q=3.267$, $r=3.303$, $s=3.314$, $t=3.387$, $u=3.451$, $v=3.457$, $w=3.410$, $x=3.467$.
(Source: R. M. Williams and S. C. Wallwork, *Acta Cryst.*, **23**, 448 (1967). Reproduced by kind permission of the International Union of Crystallography, England.)

2.3.2 Infrared spectra of EDA complexes

The infrared spectra of weak EDA complexes are similar to a superposition of the spectra of the uncomplexed components, although in hydrocarbon–quinone complexes, slight red shifts of the carbonyl and C=C double bond peaks of quinone have been reported.[37] These have been explained as being due to slight charge donation from the aromatic hydrocarbons into an antibonding orbital of the quinone, causing a slight weakening of the carbonyl and C=C bonds.

Complexes involving very strong donors and acceptors appear to be almost wholly dative, i.e. ionic, in the ground state and the infrared spectra are simply the sum of the spectra of the corresponding ions and/or ion pairs. Examples of this category are the complexes of phenothiazine with iodine[38] and of *N*,*N*,*N′*,*N′*-tetramethyl-*p*-phenylenediamine (TMPD) with *p*-chloranil.[39] Fig. 2.11 shows the infrared spectrum of TMPD–chloranil complex in the solid state. Although the carbonyl band of free chloranil

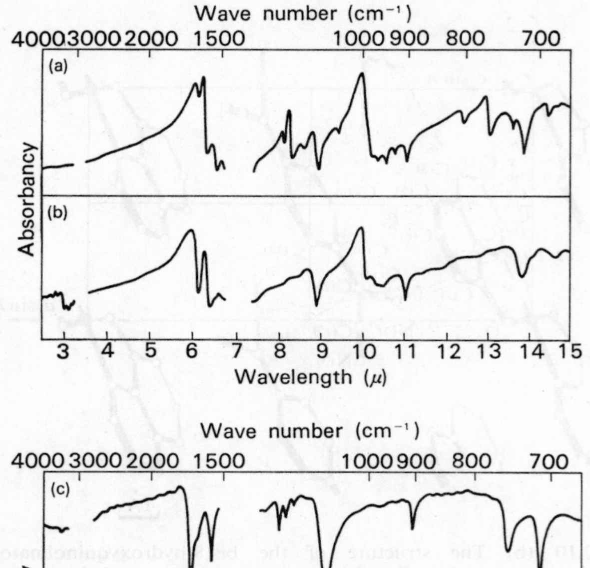

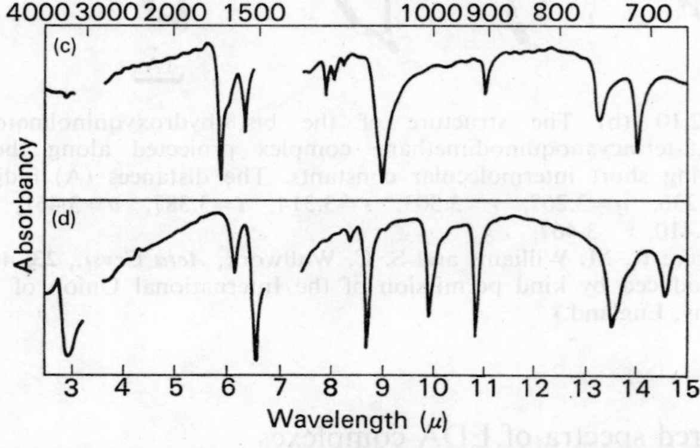

Fig. 2.11 Infrared spectra of (a) TMPD–*p*-chloranil (KBr disk), (b) durendiamine–*p*-chloranil, (c) *p*-chloranil, and (d) *p*-chloranil^{-}Li^{+} salt. (Source: Y. Matsunaga, *J. Chem. Phys.*, **41**, 1609 (1964). Reproduced by kind permission of the American Institute of Physics, U.S.A.)

lies at 1695 cm^{-1}, the complex of chloranil with TMPD has the carbonyl band at 1580 cm^{-1}, which is similar to that of *p*-chloranil^{-}Li^{+} ionic salt prepared independently. The TMPD–chloranil complex exhibits a strong esr signal at $g = 2.005$ as a semiquinone anion radical. The carbonyl bands of chloranil complexed with different organic donors shifts to longer wave number from that in the free molecule according to the extent of charge transfer in complexing.

Plyler and Mulliken have found[40] a shifted fundamental for iodine in benzene, and, in addition, noted a greatly shifted band (from 213 cm^{-1} to 184 cm^{-1}) for iodine in pyridine diluted with *n*-hexane. It is likely that the vibrational band of the trimethylamine–iodine complex found by Nagakura *et al.*[41] at 185 cm^{-1} arises from a shifted fundamental of iodine. The intensity of the acceptor bands has been explained in terms of enhancement of the dipole moment of the complex as a result of stretching. Some other infrared data have been reviewed by McGlynn.[42]

Table 2.7 presents the shifted fundamentals of complexed halogen molecules, and the extent of dative structure calculated from infrared spectroscopic data and dipole-moment measurements.[43] A good correlation has been observed between the amount of infrared band shifting and the extent of charge transfer in complexing.

In the case of complexed organic compounds, C=O stretching vibra-

TABLE 2.7

(a) The shifted fundamentals of halogen molecules in complexing ($\Delta\nu$) and the degrees of dative structure of the complexes (b^2+abS_{01})†

Complex	ν_0(cm^{-1})	$\Delta\nu$(cm^{-1})	b^2+abS_{01}	a	b
Bz–I_2	207	2^d	0.02	0.99	0.10
Bz–Br_2	312	7	0.04	0.97	0.17
Bz–Cl_2	541	16	0.06	0.96	0.20
Bz–ICl	375	20	0.11	0.93	0.28
Tol–ICl	375	19	0.10	0.94	0.27
p-Xy–ICl	375	21	0.11	0.93	0.28
Py–ICl	375	83	0.30	0.76	0.41
ICl_2^-	375	108	0.57	—	—
Py–I_2	207	24	0.29	0.78	0.43
Me_3N–I_2	207	22	{0.61 {0.41	—	—

† For small frequency shifts in weak complexes, $\frac{2\Delta\nu}{\nu}\equiv\frac{2(\nu_0-\nu)}{\nu}\simeq\frac{\Delta k}{k}$, where Δk denotes the change in force constant from the free molecule (e.g. I_2, ICl, Cl_2, etc.) to the complex. According to the Mulliken theory, approximately $(b^2+abS_{01})\simeq\frac{\Delta k}{k}$; thus, $(b^2+abS_{01})\simeq\frac{8}{3}\frac{\Delta\nu}{\nu}$.

(b) Comparison of (b^2+abS_{01}) values from infrared shifts ($\Delta\nu$) and dipole moment shifts (μ) with those from dipole moments

Complex	μ_1	b^2+abS_{01} from $\Delta\nu/\nu$	b^2+abS_{01} from μ
Bz–I_2	24.0	0.02	0.075
Py–I_2	17.7	0.29	0.25
Et_3N–I_2	17.7	(0.4)	0.28–0.35
Me_3N–I_2	17.7	0.41	0.33

TABLE 2.8 The shifted fundamentals (stretching and out-of-plane vibrations of organic compounds complexed with iodine

Electron donors	Fundamental vibration		ν_D	(cm⁻¹) (complex) ν_C	$\Delta\nu$	Ref.
Acetone	C=O	str. vib.	1716	1700	−16	(a, b)
	C–C=O	out-of-plane	529	534	+5	(b)
Ethyl ether	C–O	str. vib.	1118	1098	−20	(a, b)
Methyl acetate	C=O	str. vib.	1749	1724	−25	(b)
	C–O	str. vib.	1241	1258	+17	(b)
Dimethylformamide	C=O	str. vib.	1662	1619	−43	(c)
	C–N	str. vib.	1460	1540	+90	(c)

(a) G. L. Glusker, H. W. Thompson and R. S. Mulliken, *J. Chem. Phys.*, **21**, 1407 (1953).
(b) H. Yamada and K. Kozima, *J. Chem. Soc.*, **82**, 1543 (1960).
(c) C. D. Schmulback and R. S. Drago, *ibidy*, **82**, 4484 (1960).

tions of ketones and ethers decrease by about 20 cm^{-1} on complexing with iodine, whereas the out-of-plane vibrations of both bonds increase. Generally, the stretching vibrations of the complexed functional groups of the donor molecules decrease in wave number (red shift), whereas those of the out-of-plane vibrations increase (blue shift). According to the characteristic patterns of the shifted infrared bands, one can estimate the functional groups involved in complexing. For instance, the C=O stretching vibration of methyl acetate complexed with iodine shows a red shift, but that of C–O shows a blue shift, implying that iodine interacts with the carbonyl group of methyl acetate rather than with the C–O group. Similarly, in the *N*,*N*-dimethylformamide (DMF)–I_2 complex, the C=O stretching vibration of DMF has a red shift, whereas that of C–N has a blue shift (Table 2.8), implying that iodine might show charge-transfer binding with the carbonyl group of DMF in solution.

Other possible effects of complexing on the infrared spectra include alteration of the selection rules due to the lowering of symmetry of the complexed donor or acceptor compared to the uncomplexed molecules. D'Or *et al.* successfully demonstrated[44] that the infrared-inactive fundamentals of chlorine and bromine appeared in the infrared spectra of solutions of these halogens in aromatic hydrocarbons as electron donors.

The addition of an electron to an aromatic ketone by means of alkali metal reduction (as a strong electron donor) results in the formation of a ketyl radical anion. The addition of an electron to an aromatic hydrocarbon such as naphthalene or anthracene results in the formation of an anion radical, in which the electron goes into the lowest antibonding orbital and is unequally delocalized over the entire aromatic system. It remains to establish a direct correlation between known electron density and the shifts observed in the transformation of ketones or aromatic hydrocarbons to ketyl or aryl radical anions. For the aryl case, it is somewhat more satisfactory to conceive of the correlation in terms of π-bond energy changes rather than the bond orders. In addition, the unpaired electron is to some extent delocalized over the system of aromatic rings, and resonance energy changes in the rings are far more suitable for consideration, especially in hydrocarbon anion radicals. Within aryl ring systems, the unequally distributed unpaired electron causes bond-order changes and Coulombic distortion in the C–H bonds. These effects should result in some changes in C–H out-of-plane vibrations, which occur in the frequency range of 600–900 cm^{-1}.[45)]

Sodium and potassium metal-reduced aromatic ketyl compounds and aromatic hydrocarbons have been observed in the infrared region. The markedly shifted fundamentals of benzophenone and biphenyl complexed with sodium are shown in Fig. 2.12(a) and (b), where each spectrum is compared in reduced form with that of the free molecule in solution. Assignments of ketyl and aromatic radicals were made by following the gradual disappearance and appearance of peaks as progressive metal reduction was applied to the dimethoxyethane solution. The results are summarized in Table 2.9, where the infrared shifts of different aromatic ketones and hydrocarbons to radical anion species are presented. A similar order of infrared shifts has been observed for Co-phthalocyanine–sodium EDA complexes. Co-phthalocyanine reduced with sodium ($CoPc^{2-}$ $2Na^+$) has out-of-plane vibrations shifted to 770 and 730 cm^{-1} from 711 and 695 cm^{-1} in neutral Co-phthalocyanine. This shift might be due to partial electron donation into the aryl-π-porphin ligand of phthalocyanine, where one electron is localized in the *d* orbital of the Co^{2+} central ion and

TABLE 2.9 Infrared shifts of different aromatic ketones and hydrocarbons and the corresponding radical anions in DME

	Neutral molecule (cm^{-1})		Observed radical assignments (cm^{-1})	
Benzophenone	C=O	1664	ketyl	1554
	Ar	1598	Ar	1581
Fluorenone	C=O	1720	ketyl	1540
	Ar(doublet)	1612, 1602	Ar	1580
Biphenyl	Ar	1600	Ar	1568
Naphthalene	Ar	1600	Ar	1587, 1521

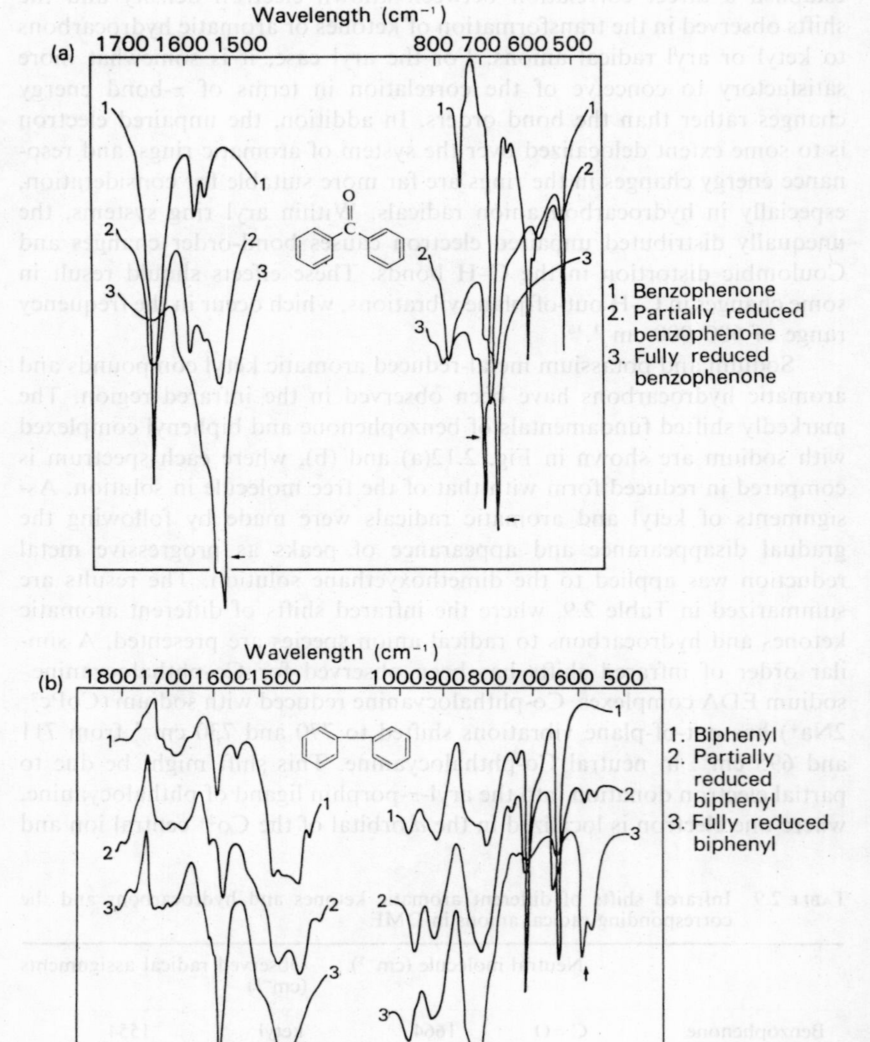

Fig. 2.12 Infrared spectra of reduced benzophenone (a) and biphenyl (b) in dimethoxyethane in comparison with those of the free molecules.

the other is delocalized over the entire π system of the phthalocyanine ligand.[46]

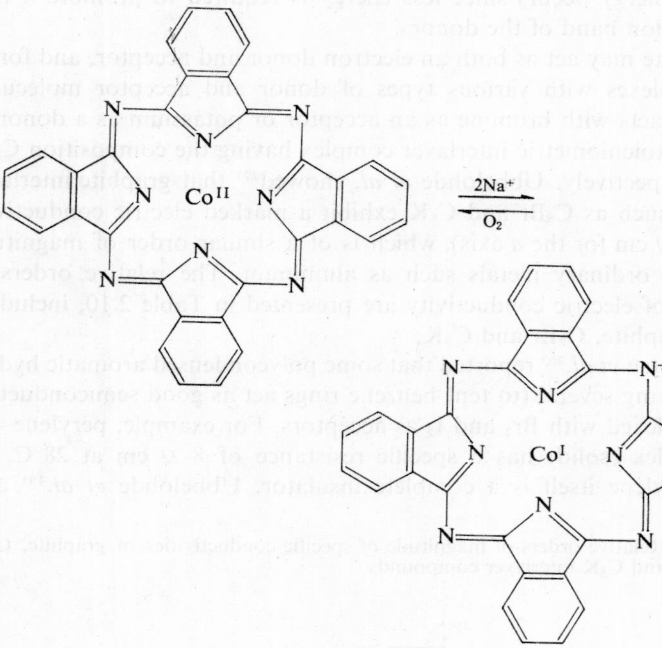

2.3.3 Electric conductivity and magnetic properties of EDA complexes

Although most aromatic hydrocarbon crystals are electric insulators, graphite with its two-dimensional polycondensed structure is a good conductor. The specific resistance of graphite has been measured at $6 \times 10^{-5}\ \Omega$ cm for the *a* axis and 0.5 Ω cm for the *c* axis, perpendicular to the graphite molecular plane. Various studies have been reported, in which charge-transfer complexing has been inferred from marked changes in the conductivity of graphite produced by introducing donor or acceptor components. The semiconductivity of organic charge-transfer complexes is very different from those of the individual components. The complexes have conductivities which are several orders of magnitude greater than those of the components. There is always a concomitant decrease in the activation energy. These conductivity changes on complexing can be explained in terms of the removal of an electron from the donor to the acceptor, such as graphite or polycondensed aromatic hydrocarbons, leaving a "hole" into which other electrons may flow. The decrease in activation energy occurs since less energy is required to promote it into the conduction band of the donors.

Graphite may act as both an electron donor and acceptor, and forms EDA complexes with various types of donor and acceptor molecules. Graphite reacts with bromine as an acceptor or potassium as a donor to produce a stoichiometric interlayer complex having the composition C_8Br or C_8K, respectively. Ubbelohde *et al.* showed[47] that graphite interlayer complexes such as C_8Br and C_8K exhibit a marked electric conductivity (10^{-4}–10^{-5} Ω cm for the *a* axis), which is of a similar order of magnitude to those of ordinary metals such as aluminum. The relative orders of magnitude of electric conductivity are presented in Table 2.10, including those of graphite, C_8Br and C_8K.

Akamatsu *et al.*[48] reported that some polycondensed aromatic hydrocarbons having several (to ten) benzene rings act as good semiconductors when complexed with Br_2 and I_2 as acceptors. For example, perylene–Br_2 (1:2) complex (solid) has a specific resistance of 8 Ω cm at 28°C, although perylene itself is a complete insulator. Ubbelohde *et al.*[49] also

TABLE 2.10 Relative orders of magnitude of specific conductivities of graphite, C_8Br and C_8K interlayer compounds

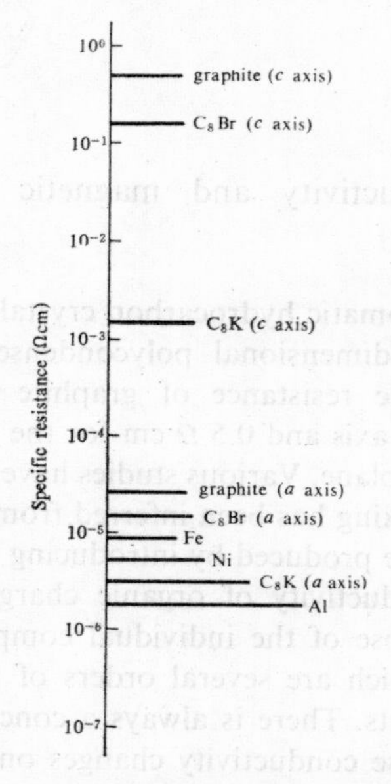

(After Y. Matsunaga, *Kagaku Zōkan* (Japanese), **48**, p. 159, Kagaku Dojin, 1971.)

reported semiconductive behavior in the solid anthracene–sodium complex (anthracene–$Na^{2.02}$) with a specific resistance of 10^{10} Ω cm at 28°C *in vacuo*.

Although both anthracene and perylene are similar electric insulators, they play different roles in complexing with electron donors or acceptors. Anthracene acts as an acceptor with sodium, but perylene as a donor with bromine to give a good semiconductor. In many cases, polycondensed aromatic hydrocarbons act as good donors rather than as acceptors to form EDA complexes.

In Table 2.11 the specific resistance and the activation energies are summarized for complex solids between various condensed aromatic and heteroaromatic hydrocarbons with halogens. Slight changes in the

TABLE 2.11 Electric conductivities of the solid complexes between aromatic hydrocarbons and halogens

Complex (molar ratio)	Relative resistance (Ω cm)	E_a (eV)	Ref.
Perylene–Br_2 (1:2)	8	0.065	(a, b)
Perylene–I_2 (1:3)	10	0.03	(b)
Perrylene–I_2 (2:3)	8	0.019	(b)
Pyrene–I_2 (1:2)	75	0.14	(c)
Pyranthrene–I_2 (1:2)	20	0.045	(d)
Violanthrene–I_2 (1:2)	13	0.075	(b)
Naphthacene–I_2 (1:0.88)	10^{10}	0.54	(e)
1,2-Benzanthracene–I_2 (1:0.87)	10^{12}	0.90	(e)
β-Carotene–I_2 (2:3)	10^8	0.55	(f)
Pyranthrene–Br_2 (1:1.7)	10^2	0.10	(b)
Violanthrene–Br_2 (1:2.3)	66	0.10	(b)
Perylene–ICl (1:1.9)	10^4	0.17	(g)

(After Y. Matsunaga, *Yūki Handōtai* (Japanese), p. 182, Kyoritsu Shuppan, 1966.)

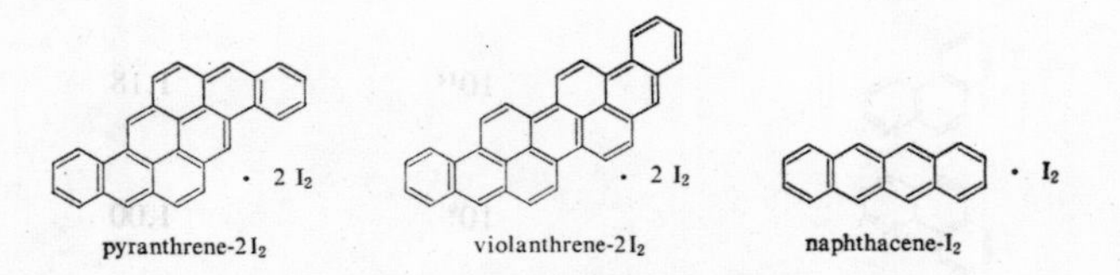

(a) H. Akamatsu, H. Inokuchi and Y. Matsunaga, *Nature*, **173**, 168 (1954).
(b) H. Akamatsu, H. Inokuchi and Y. Matsunaga, *Bull. Chem. Soc. Japan*, **29**, 213 (1956).
(c) J. Kommandeur and F. R. Hall, *J. Chem. Phys.*, **34**, 129 (1961).
(d) T. Uchida and H. Akamatsu, *Bull. Chem. Soc. Japan*, **35**, 981 (1962).
(e) G. C. Martin and A. R. Ubbelohde, *J. Chem. Soc.*, **1961**, 4948.
(f) C. M. Huggins and O. H. LeBlanc, *Nature*, **186**, 552 (1960).
(g) M. S. Frant and R. Eiss, *J. Electrochem. Soc.*, **110**, 769 (1963).

TABLE 2.12 Specific resistances and activation energies of solid complexes of aromatic (or heteroaromatic) compounds with alkali metals

Complexes	Specific resistance (Ω cm)	E_a (eV)	Ref.
$Li_{1.16}$ 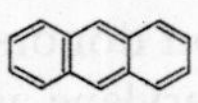	10^{11}	1.34	(a)
$K_{1.18}$ 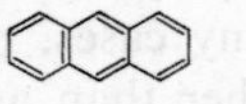	10^{11}	1.10	(a)
$Na_{2.08}$ 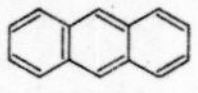	10^{8}	0.69	(b)
$Na_{1.8}$ 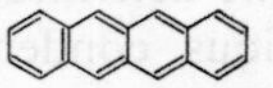	10^{7}	0.32	(c)
$Na_{1.02}$ 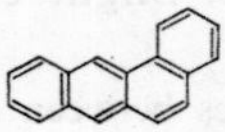	10^{9}	0.667	(c)
$Na_{2.37}$ 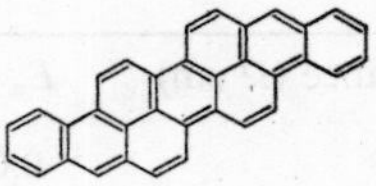	61	0.048	(d)
$K_{4.35}$ ″	27	0.030	(d)
$Na_{1.97}$ 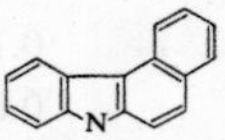	10^{8}	0.32	(c)
$Na_{0.93}$ 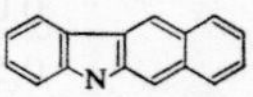	10^{8}	0.87	(c)
$Na_{1.99}$ 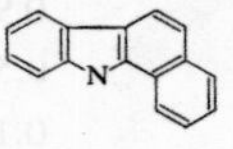	10^{10}	0.50	(c)
$Na_{1.07}$ 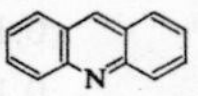	10^{14}	1.99	(e)
$Na_{1.49}$ 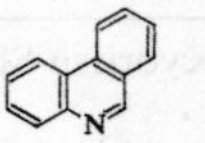	10^{8}	0.35	(e)
$Na_{1.60}$ 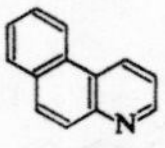	10^{11}	1.18	(e)
$Na_{2.09}$ 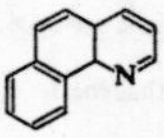	10^{9}	1.00	(e)

(After K. Matsunaga, *Yūki Handōtai* (Japanese), p. 185, Kyoritsu Shuppan, 1966.)

(a) W. A. Holmes-Walker and A. R. Ubbelohde, *J. Chem. Soc.*, **1954**, 720.
(b) J. P. V. Gracey and A. R. Ubbelohde, *ibid.*, **1955**, 4089.
(c) G. C. Martin and A. R. Ubbelohde, *ibid.*, **1961**, 4948.
(d) N. D. Parkyns and A. R. Ubbelohde, *ibid.*, **1961**, 2110.
(e) S. Slough and A. R. Ubbelohde, *ibid.*, **1957**, 982.

compositions of the complexes cause marked changes of the electric conductivity.

Table 2.12 includes experimental data for the electric conductivity and activation energy of EDA complexes between aromatic (and heteroaromatic) hydrocarbons and alkali metals (Na and K) in the solid state. Isoviolanthrene–alkali metal complexes (1:2–1:4) are fairly good semiconductors *in vacuo*. These complexes are very sensitive to air and water vapor.

Most organic complexes with high conductivities have been reported as ionic EDA complexes consisting of radical anion and cation salts. Delocalized electrons or "holes" in complex solids act as charge carriers. Recently Du Pont chemists have developed tetracyanoquinodimethane (TCNQ) radical ion salts with metal or ammonium cations as excellent conductors,[50] almost like synthetic metals. The specific conductivities of these complex salts are the largest so far known for organic materials (Table 2.13, 2.14).

β-Carotene displays strong esr signals on mixing with iodine.[51] This mixture also behaves as a semiconductor with a low activation energy and high conductivity typical of charge-transfer complexes. The temperature dependency of the conductivity is quite different from that of the esr signal,

TABLE 2.13 Specific resistance of some molecular CT complexes of tetracyanoquinodimethane (TCNQ) as an acceptor with various organic donors

Donor	Specific resistance (Ω cm)	Donor	Specific resistance (Ω cm)
$NH_2-C_6H_4-NH_2$	10^3	$NH_2-C_6(CH_3)_4-NH_2$	10^{10}
$NH_2-C_6H_3(CH_3)-NH_2$	10^5	$NH_2-C_{10}H_6-NH_2$	10^9
$NH_2-C_6H_4-N(CH_3)_2$	10^9	$C_{14}H_{10}$	10^{11}
$CH_3(H)N-C_6H_4-N(H)CH_3$	10^4	$C_{16}H_{10}$	10^{12}
$(CH_3)_2N-C_6H_4-N(CH_3)_2$	10^6		

(Source: L. R. Melby, R. J. Harder, W. R. Hertler, W. Mahler, R. E. Benson and W. E. Mochel, *J. Am. Chem. Soc.*, **84**, 3374 (1962). Reproduced by kind permission of the American Chemical Society, U.S.A.)

TABLE 2.14 Electrical properties of solid complexes between *p*-phenylene diamine derivatives and iodine

Complex	Specific resistance (Ω cm)	E_a(eV)	Ref.
$NH_2-C_6H_4-NH_2 \cdot 0.82\, I_2$	1.7×10^5	0.41	(a)
$NH_2-C_6H_4-C_6H_4-NH_2 \cdot 1.26\, I_2$	2.2	0.19	(b)
$C_6H_5-(NH-C_6H_4)_2-C_6H_5 \cdot 1.2\, I_2$	2.9	0.06	(c)
$C_6H_5-(NH-C_6H_4)-C_6H_5 \cdot 1.5\, I_2$	1.0	0.05	(c)
$C_6H_5-(NH-C_6H_4)_6-C_6H_5 \cdot 1.05\, I_2$	2.6	0.042	(c)
" $\cdot 2\, I_2$	1.1	0.021	(c)
" $\cdot 3\, I_2$	0.8	0.031	(c)
$NH_2-C_6H_4-(NH-C_6H_4)_2-C_6H_4-NH_2 \cdot 1.1\, I_2$	1.8	0.03	(c)
$(CH_3)_2N-C_6H_4-(NH-C_6H_4)_2-C_6H_4-N(CH_3)_2 \cdot I_2$	1000	0.24	(c)

(a) S. Nishizaki and H. Kusakawa, *Bull. Chem. Soc. Japan*, **36**, 1681 (1963).
(b) H. Kusakawa and S. Nishizaki, *ibid.*, **38**, 313 (1965).
(c) V. Hádek, P. Zach, K. Ulbert and J. Honzl, *Coll. Czech. Chem. Commun.*, **34**, 3139 (1969).

which means either that the esr intensity is not a measure of the number of conduction carriers or that the determining factor in the temperature dependency of the conductivity is the mobility of the carriers rather than their number. The following reaction scheme might explain the spectroscopic results.

β-carotene

$$\beta\text{-carotene} + 2I_2 \rightleftharpoons (\text{carotene}\cdots\text{I})^+ + I_3^-$$

Carotene forms a charge-transfer complex with I^+ ion. The new absorption band at 1000 nm is thought to arise from charge-transfer transition between

TABLE 2.15 Electrical properties of TCNQ salts with various quaternary alkylammonium cations

Complex salt	Specific resistance (Ω cm)	E_a	Ref.
NH_4^+ $TCNQ^{\overline{\cdot}}$	4.3×10^4	0.22	(a)
Me_4N^+ $TCNQ^{\overline{\cdot}}$	3.4×10^5	0.22	(a)
Me_4N^+ $(TCNQ)_2$	1.1×10^6	0.37	(a)
Et_4N^+ $(TCNQ)^{\overline{\cdot}}$	1.0×10^6	0.28	(a)
Et_4N^+ $(TCNQ)_2^{\overline{\cdot}}$	3.3×10^4	0.34	(a)
n-Pr_4N^+ $(TCNQ)^{\overline{\cdot}}$	1.6×10^9	0.87	(a)
n-Pr_4N^+ $(TCNQ)_2^{\overline{\cdot}}$	9.9×10^3	0.33	(a)
n-Bu_4N^+ $(TCNQ)^{\overline{\cdot}}$	4.8×10^{10}	0.88	(a)
n-Bu_4N^+ $(TCNQ)_2^{\overline{\cdot}}$	9.7×10	0.21	(a)
$(n$-$C_5H_{11})_4N^+$ $(TCNQ)_2^{\overline{\cdot}}$	1.0×10^7	0.75	(a)
$(n$-$C_6H_{13})_4N^+$ $(TCNQ)_2^{\overline{\cdot}}$	4.6×10^8	0.93	(a)
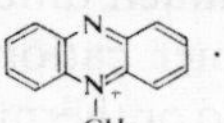 N, N^+, CH_3 $(TCNQ)_2^{\overline{\cdot}}$	1.4	—	(b)
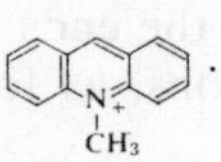 N^+, CH_3 $\cdot$ $(TCNQ)_2^{\overline{\cdot}}$	0.8	0.025	(c)
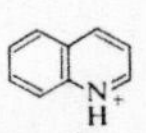 N^+, H $\cdot$ $(TCNQ)_2^{\overline{\cdot}}$	0.5	0.025	(c)

(a) H. Kusakawa and K. Akashi, *Bull. Chem. Soc. Japan*, **42**, 263 (1969).
(b) L. R. Melby, *Can. J. Chem.*, **43**, 1448 (1965).
(c) P. Dupuis and Néel, *Compt. Rend. Sér. C.*, **265**, 688, 777 (1967).

the complex in its ground state (carotene$^-$...I$^+$) and the excited complex (carotene$^+$....I). It has been demonstrated[52] that the conductivities of TCNQ salts depend not only upon the stoichiometry of the complexes, but also upon the size of the alkylammonium cation, as shown in Table 2.15. The conductivity of 1:1 TCNQ complex with different quaternary alkyl ammonium cations increases with the size of the alkyl groups.[52] In contrast to this, that of the 1:2 complex salts has a maximum value with tetra-n-butyl ammonium; 97 Ω cm^{-1} at 28°C. The change of conductivity with the structure of the ammonium cation in TCNQ salts is presumably due to the geometrical requirements for maximum electronic interactions between $TCNQ^{\overline{\cdot}}$ and neutral TCNQ in the crystal lattice of the complex salts.

The electric conductivity of the deeply colored complex solids between polymer donors such as poly(ethynylnaphthalene), poly(vinylanthracene), poly(vinylpyridine) and iodine have been reported.[53] By using N'-alkylpoly(vinylpyridinium) as a polymer cation, various TCNQ salts have been prepared[54] by reaction with Li^+TCNQ^- solution, as follows:

$-(CH_2-CH)_n-$ (pyridinium, N^+–H) + $Li^+ TCNQ^-$ ⟶ $-(CH_2-CH)_n-$ (pyridinium, N^+–H) $(TCNQ)_p^- (TCNQ)_q$

$-(CH_2-CH)_n-$ (pyridinium, N^+–R) + $Li^+ TCNQ^-$ ⟶ $-(CH_2-CH)-$ (pyridinium, N^+–R) $(TCNQ)^-$

The conductivity of TCNQ salts with polymer cations is not much different from that of TCNQ salts with the corresponding momomer cations.

A series of cyanine dyes is known to be polarizable by photo or thermal excitation. Electrons (or charge) may be transported between the ends of the molecule through conjugate systems by small perturbations, as formally described in the following transition:

(quinolinium, N^+–Et)–CH=CH–CH=(quinoline, N–Et) ⇌ (quinoline, N–Et)=CH–CH=CH–(quinolinium, N^+–Et)

Many TCNQ salts have been synthesized[55] with cyanine compounds as counter cations. These showed marked electrical conductivity changes depending on the size of the cyanine compounds. Since the polarizability may depend upon the size of the cyanine compounds, larger cyanine compounds form better conductors in combination with TCNQ anions, as shown in Table 2.16. The physical properties of the carriers in complex solids have been measured in terms of the thermovoltage, e.g. "Hall coefficient" measurements. It is known that electrons act as carriers in conduction through complex solids such as perylene–DDQ (2,3-dichloro-5,6-dicyano-*p*-quinone),[56] $Cs_3(TCNQ)_2^-$ and $Li^+(TCNQ)^-$,[57] whereas

TABLE 2.16 Electrical properties of TCNQ–cyanine complexes

TCNQ–cyanine complex (mole ratio)		Specific resistance (Ω cm)	E_a (eV)	Ref.
N+ –CH= N (Et, Et)	(1:1)	10^9	—	(a)
	(1:2)	3.6	0.077	(b)
S, N+ –CH=CH–CH= S, N+ (Et, Et)	1:2	5.0	0.056	(c)
S, N –CH=CH–CH= S, N+ (Et, Et)	1:2	6.7	0.067	(b)
S, N+ –CH= S, N+ (Et, Et)	1:2	150	0.21	(b)

(a) J. H. Lupinski, K. R. Walter and L. H. Vogt Jr., *Mol. Cryst.*, 3, 241 (1967).
(b) B. H. Klanderman and D. C. Hoesterey, *J. Chem. Phys.*, **51**, 377 (1969).
(c) E. B. Yagubskii, M. L. Khidekel and I. F. Shchegolev, *Zh. Obshch. Khim.*, **38**, 992 (1968).

holes act as carriers in complexes of perylene, violanthrene, and acridine with iodine,[58] *p*-phenylenediamine–DDQ, perylene–$RuCl_3$,[59] Cs(TCNQ), and $Ba(TCNQ)_2$. It is interesting to note that electrons behave as carriers in perylene–$FeCl_3$ with an excess of $FeCl_3$, but holes do so in the presence of excess perylene.

2.3.4 Paramagnetism in complexes

The complexes with high electric conductivity display strong esr signals and paramagnetism in the ground state. The explanation[60] for paramagnetism in complexes is based on a consideration of the ionic state of an adduct derived from components with filled electron shells. The free ions $A\cdot^-$ and $D\cdot^+$ each have an unpaired electron and are in the doublet state. In a complex these electrons may interact to provide a singlet and a triplet state. Generally the paramagnetic triplet state is energetically too high to be populated at room temperature. Complexes which are formed from donors of low ionization potential and acceptors of high electron affinity may, in contrast to most adducts, be significantly ionic in character. Under these circumstances the triplet state may not lie far above the ground state, in which case it may be sufficiently populated at room temperature so that the complex is measurably paramagnetic. Factors which should lead to increased population of this level are (1) increased ease of reduction of the acceptor and (2) increased bulkiness of substituents on the donor and

TABLE 2.17 Relative strengths of esr absorption per molecule of complex at 90°K

Acceptor	*p*-Phenylenediamine complexes	*N,N,N′,N′*-tetramethyl-*p*-phenylenediamine complexes
p-Benzoquinone	0.0	0.1
p-Chloranil	0.2	0.2
p-Bromanil	0.3	2.0
p-Iodanil	7.0	20.0
o-Bromanil	3.0	40.0

(Source: D. Bijl, H. Kainer and A. C. Rose-Innes, *J. Chem. Phys.*, **30**, 767 (1959). Reproduced by kind permission of the American Institute of Physics, U.S.A.)

acceptor nuclei (with a corresponding decrease in overlap of the orbitals of $A\cdot^-$ and $D\cdot^+$ which contain the unpaired electrons). Both of these factors may contribute to the observed increases in the relative strengths of esr absorption per molecule of *p*-phenylenediamine and *N,N,N′,N′*-tetramethyl-*p*-phenylenediamine complexes with changes in acceptor, as outlined in Table 2.17. In assigning these relative strengths a radical having a simple doublet magnetic ground state has been used as a reference substance.

In several cases the spin concentrations of these solid amino–quinone adducts follow the Curie law; that is, they vary inversely with the absolute temperature. The esr signals of certain of the solid adducts of tetrahaloquinones with *p*-phenylenediamine and *N,N,N′,N′*-tetramethyl-*p*-phenylenediamine have been resolved, and the spectral lines can be identified with the ion-radicals, D^+ and A^-. Apparently there is little interaction between these ions in the quinone–amine type paramagnetic complexes.

The ionic character of an adduct of this type diminishes as the solvent becomes less polar. Solutions of the tetraphenylenediamine-halogenated quinone complexes are deep blue in acetonitrile or water–methanol mixtures, but they are almost colorless in benzene or dioxane.

The solid iodine adducts of a variety of polycyclic aromatic hydrocarbons[61] including perylene (a), pyrene (b), and violanthrene (c) display

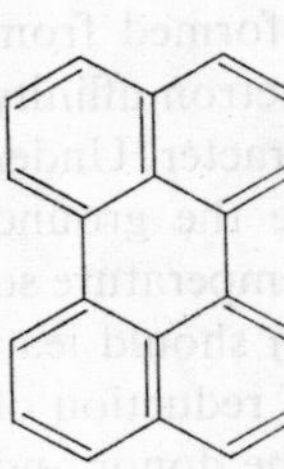

(a) 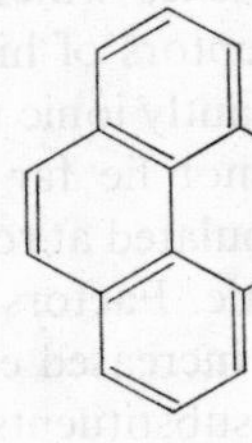 (b) 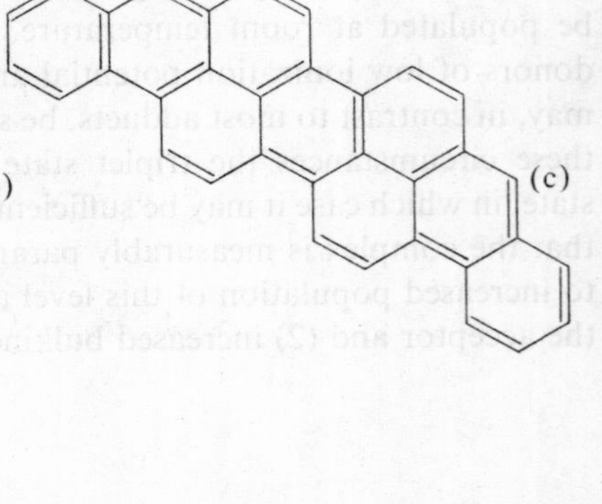 (c)

esr absorption. Bromine and iodine adducts of certain polyphenylethylenes and substituted diphenyls are also reported to be paramagnetic.[62]

Both the relative spin concentrations and the specific resistivities of solid iodine complexes of perylene and pyrene show exponential temperature dependence according to the equation

$$X = X_0 \, e^{-E/kT}$$

Actually for each of these complexes the activation energies, E, for conduction and for spin concentration are essentially the same over wide temperature ranges. This is illustrated in Fig. 2.13 in which values of log

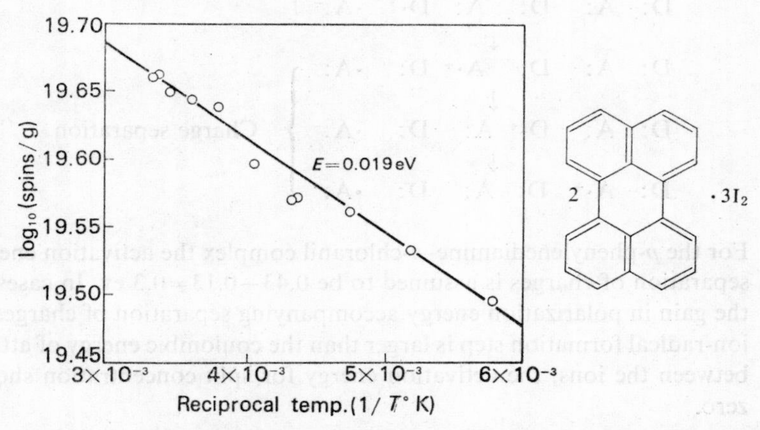

Fig. 2.13 Temperature dependence of the spin concentration for $(\text{perylene})_2 \cdot 3I_2$ complex. The line is based on the observed E value of 0.019 eV for conductivity of the complex; the points are based on measured spin concentrations.
(Source: L. S. Singer and J. K. Kommandeur, *J. Chem. Phys.*, **34**, 133 (1961). Reproduced by kind permission of the American Institute of Physics, U.S.A.)

X for the perylene adduct are plotted against $1/T$. The slopes of such plots are used in calculating the activation energies. A similar correlation is reported for the iodine–violanthrene complex.[63] It may be reasonably concluded that in these complexes the unpaired spins are associated with the charge carriers in conduction.

In cases in which relative spin concentrations vary with temperature according to the above equation, the activation energy for esr absorption corresponds to the energy required to promote a singlet-triplet transition or to promote the transition from a nonionic ground state to an excited state composed of two independent doublets.

Ottenberg, Hoffman, and Osiecki[64] have discussed a possible mechanism of conduction of the quinone–amine complexes which have E values for conduction which are zero or substantially less than those for spin concentration. The first step in the scheme is the formation of ion-radicals, with an activation energy of 0.13 ev in the case of the *p*-phenylenediamine–chloranil complex. This is followed by charge separation by transfer of electrons as indicated in a simplified scheme which is based on the assumptions that the donors (D:) and acceptors (A:) are alternately stacked and that the acceptor is amphoteric in character.

$$
\begin{array}{l}
\left.\begin{array}{llllll}
D: & A: & D: & A: & D: & A: \\
 & & & \downarrow & & \\
D: & A: & D: & A: & D\cdot^{+} & \cdot A:^{-}
\end{array}\right\} \text{Ion-radical formation} \\
\qquad\qquad\qquad\downarrow \\
\left.\begin{array}{llllll}
D: & A: & D: & A\cdot^{+} & D: & \cdot A:^{-} \\
 & & & \downarrow & & \\
D: & A: & D\cdot^{+} & A: & D: & \cdot A:^{-} \\
 & & & \downarrow & & \\
D: & A\cdot^{+} & D: & A: & D: & \cdot A:^{-}
\end{array}\right\} \text{Charge separation}
\end{array}
$$

For the *p*-phenylenediamine–*p*-chloranil complex the activation energy for separation of charges is assumed to be $0.43-0.13=0.3$ ev. In cases where the gain in polarization energy accompanying separation of charges in the ion-radical formation step is larger than the coulombic energy of attraction between the ions, the activation energy for spin concentration should be zero.

REFERENCES

1. I. Isenberg and A. Szent-Györgyi, *Proc. Natl. Acad. Sci. U.S.A.*, **44**, 857 (1958); **45**, 1229 (1959).
2. W. J. James, D. French and R. E. Rundle, *Acta Cryst.*, **12**, 358 (1959).
3. C. E. Castro, L. J. Andrews and R. M. Keefer, *J. Am. Chem. Soc.*, **80**, 2322 (1958).
4. L. H. Klemn and G. W. Sprague, *J. Org. Chem.*, **19**, 1464 (1954).
5. T. W. Nakagawa, L. J. Andrews and R. M. Keefer, *J. Am. Chem. Soc.*, **82**, 269 (1960).
6. K. N. Trueblood and H. J. Lucas, *ibid.*, **74**, 1338 (1952).
7. M. S. Newman and W. B. Lutz, *ibid.*, **78**, 2469 (1956).
8. M. Green and R. F. Hudson, *Proc. Chem. Soc.*, **1957**, 323.

9. M. S. Newman and D. Lednicer, *J. Am. Chem. Soc.*, **78**, 4765 (1956).
10. L. H. Klemn, *J. Chromat.*, **3**, 364 (1960).
11. R. J. Cvetanovič, F. J. Duncan, W. E. Falconer and W. F. Sunder, *J. Am. Chem. Soc.*, **88**, 1602 (1966); W. E. Falconer and R. J. Cvetanovič, *J. Chromat.*, **27**, 20 (1967).
12. A. R. Cooper, C. W. P. Crowne and P. G. Farrell, *Trans. Faraday Soc.*, **62**, 2726 (1966); **63**, 447 (1967).
13. A. R. Cooper, C. W. P. Crowne and P. G. Farrell, *J. Chromat.*, **27**, 362 (1967).
14(a). M. Frank-Neumann and P. Jössang, *ibid.*, **14**, 280 (1964).
(b). R. O. C. Norman, *Proc. Chem. Soc.*, **1958**, 151.
15. R. G. Harvey and M. Halonen, *J. Chromat.*, **25**, 294 (1966).
16. H. Kessler and E. Müller, *ibid.*, **24**, 469 (1966).
17. D. B. Parihar, S. P. Sharma and K. K. Verma, *ibid.*, **31**, 120 (1967).
18. M. Godlewicz, *Nature*, **164**, 1132 (1949).
19(a). A. Berg and J. Lam, *J. Chromat.*, **16**, 157 (1964); L. H. Klemn, D. Reed and C. D. Lind, *J. Org. Chem.*, **22**, 739 (1957).
(b). N. P. Buu-Hui and P. Jacquignon, *Experientia*, **13**, 375 (1957).
20. J. T. Ayres and C. K. Mann, *Anal. Chem.*, **36**, 2185 (1964).
21. W. N. White, *J. Am. Chem. Soc.*, **81**, 2921 (1959).
22. H. A. H. Craenen, J. W. Verhoeven and J. de Boer, *Rec. Trav. Chim. Pays-Bas*, **91**, 405 (1972).
23. D. J. Cram and R. H. Bauer, *J. Am. Chem. Soc.*, **81**, 5971 (1959).
24. D. J. Cram and D. I. Wilkinson, *ibid.*, **82**, 5721 (1960).
25. O. Hassel and K. O. Stroemne, *Acta Chem. Scand.*, **12**, 1146 (1958).
26. E. E. Ferguson, *J. Chem. Phys.*, **25**, 577 (1966); **26**, 1357 (1957).
27. P. Groth and O. Hassel, *Acta Chem. Scand.*, **18**, 402 (1964).
28. O. Hassel and C. Rømming, *ibid.*, **10**, 796 (1956).
29. H. G. Smith and R. E. Rundle, *J. Am. Chem. Soc.*, **80**, 5075 (1958).
30. D. S. Brown, S. C. Wallwork and A. Wilson, *Acta Cryst.*, **17**, 168 (1964).
31. C. A. Langhott and C. J. Fritchie, *Chem. Commun.*, **1970**, 20.
32. B. Pullman and A. Pullman, *Quantum Biochemistry*, Interscience, 1963.
33. R. M. Williams and S. C. Wallwork, *Acta Cryst.*, *B*22, 897 (1967); I. Ikemoto and H. Kuroda, *ibid.*, *B*24, 383 (1968); I. Ikemoto, K. Yakushi and H. Kuroda, *ibid.*, *B*26, 800 (1970).
34(a). C. A. Bear, J. M. Waters and T. N. Waters, *Chem. Commun.*, **1970**, 702.
(b). T. T. Harding and S. C. Wallwork, *Acta Cryst.*, **8**, 787 (1955).
35(a). Y. Matsunaga, *J. Chem. Phys.*, **41**, 1609 (1964); B. Kamenar, C. K. Prout and J. D. Wright, *J. Chem. Soc.* (A), **1967**, 469.
(b). R. M. Williams and S. C. Wallwork, *Acta Cryst.*, **23**, 448 (1967).
36. H. Tsubomura and S. Nagakura, *J. Chem. Phys.*, **27**, 819 (1957).
37. G. L. Glusker and H. W. Thompson, *J. Chem. Phys.*, **21**, 1407 (1953); *J. Chem. Soc.*, **1955**, 471.
38. Y. Matsunaga, *J. Chem. Phys.*, **42**, 1982 (1969).
39. H. Kainer and W. Otting, *Chem. Ber.*, **88**, 1921 (1955).
40. E. K. Plyler and R. S. Mulliken, *J. Am. Chem. Soc.*, **81**, 823 (1959).
41. H. Yaka, J. Tanaka and S. Nagakura, *J. Mol. Spect.*, **9**, 461 (1962).
42. S. P. McGlynn, *Chem. Rev.*, **58**, 1113 (1958).
43. R. S. Mulliken and W. B. Person, *Molecular Complexes*, p. 67–68, Interscience, 1969.
44. L. D'Or, R. Alewaters and J. Collin, *Rec. Trav. Chim.*, **75**, 862 (1956).
45. D. H. Eargle Jr., *J. Chem. Soc.* (B), **1970**, 1556.
46. Y. Bansho, T. Shimura, O. Ueda, H. Takazaki and K. Takei, *J. Ind. Chem.* (*Japan*), **71**, 150 (1971).
47. F. R. M. McDonnell, R. C. Pink and A. R. Ubbelohde, *J. Chem. Soc.*, **1951**, 191.
48. H. Akamatsu, H. Inokuchi and Y. Matsunaga, *Nature*, **173**, 168 (1954).

49. W. A. Holmes-Walker and A. R. Ubbelohde, *J. Chem. Soc.*, **1954**, 720.
50. D. S. Acker, R. J. Harder, W. R. Hertler, W. Mahler, L. R. Melby, R. E. Benson and W. E. Nockel, *J. Am. Chem. Soc.*, **82**, 6408 (1960); L. R. Melby, R. J. Harder, W. R. Hertler, W. Mahler, R. E. Benson and W. E. Mochel, *ibid.*, **84**, 3374 (1962).
51. C. M. Huggins and O. H. Le Blanc, *Nature*, **186**, 552 (1960).
52. H. Kusaka and K. Akashi, *Bull. Chem. Soc. Japan*, **42**, 263 (1969); L. R. Melby, *Can. J. Chem.*, **43**, 1448 (1965).
53. H. Inoue, K. Noda, S. Takiuchi and E. Imoto, *Ind. Chem.* (*Japan*), **65**, 1286 (1962); W. Slough, *Trans. Faraday Soc.*, **58**, 2360 (1962); A. M. Hermann and A. Rembaum, *J. Polymer. Sci.*, *C***17**, 107 (1967).
54. J. H. Lupinski, K. D. Kopple and J. J. Hertz, *J. Polymer. Sci.*, *C***16**, 1561 (1967).
55. J. H. Lupinski, K. R. Walter and L. H. Vogt Jr., *Mol. Cryst.*, **3**, 241 (1967).
56. A. Ottenberg, R. L. Brandon and M. E. Browne, *Nature*, **201**, 1119 (1963).
57. W. J. Siemons, P. E. Bierstedt and R. G. Kepler, *J. Chem. Phys.*, **39**, 3523 (1963).
58. G. C. Martin and A. R. Ubbelohde, *J. Chem. Soc.*, **1961**, 4948.
59. M. S. Frant and R. Eiss, *J. Electrochem. Soc.*, **110**, 769 (1963).
60. H. Kainer and A. Überle, *Chem. Ber.*, **88**, 1147 (1955).
61. L. S. Singer and J. Kommandeur, *J. Chem. Phys.*, **34**, 133 (1961); M. Kinoshita, *Bull. Chem. Soc. Japan*, **36**, 307 (1963).
62. H. M. Buck, J. H. Lupinski and L. J. Osterhot, *Mol. Phys.*, **1**, 196 (1958).
63. H. Akamatsu, Y. Matsunaga and H. Kuroda, *Bull. Chem. Soc. Japan*, **30**, 618 (1957).
64. A. Ottenberg, C. J. Hoffman and J. Osiecki, *J. Chem. Phys.*, **38**, 1898 (1963).

CHAPTER 3

HOMOGENEOUS CATALYSIS BY EDA COMPLEXES

3.1 Chemical Reactions Involving EDA Complexes

According to the extent of charge transfer between the donor and acceptor components, they may exist as weak charge-transfer complexes or as ionic complexes (and/or σ complexes). It may be possible to make an analogy between EDA complex formation and bimolecular reactions such as substitution, rearrangement and exchange reactions among the donor and acceptor reagents.

Some theoretical and experimental studies on the possible roles of EDA complexes as intermediates have been done in aromatic electrophilic (and nucleophilic) substitution reactions. For example, it has been suggested[1] that π-complex formation of aromatic rings with the nitronium ion (NO_2^+) is the rate-determining step in the nitration of aromatic compounds on the basis of spectroscopic measurements, as shown in the following scheme for the nitration of benzene:

$$C_6H_6 + NO_2^+BF_4^- \rightleftharpoons \left[C_6H_6 \leftarrow NO_2^+ \;\; BF_4^- \right] \rightleftharpoons$$

π complex

$$\left[C_6H_6(H)(NO_2)^{+} \;\; BF_4^- \right] \rightleftharpoons C_6H_5NO_2 + H^+BF_4^-$$

σ complex

It has also been reported[2] that the two different kinds of complexes are quickly formed from a single set of reactants such as sodium ethoxide and 2,4,6-trinitroanisole. When solutions of ethoxide ion and 2,4,6-trinitroanisole were mixed at $-60°$ to $-80°C$, the solution rapidly became yellow, and this color gradually intensified to yield the covalently bonded so-called "Meisenheimer"[3] complex on raising the temperature to 0°C. By spectroscopic measurements, the yellow-colored product formed in this substitution reaction has been identified as a CT complex (π complex), which is converted to a σ complex (Meisenheimer complex) by the following substitution process:

CH_3 | O | O_2N | NO_2 | NO_2 $+ \; C_2H_5O^-$ $\underset{k_{-1}}{\overset{k_1}{\rightleftharpoons}}$ [CH_3 | O | O_2N | NO_2 | NO_2 ← $C_2H_5O^-$] $\rightleftharpoons$

CH_3–O, O–CH_2CH_3 | O_2N | NO_2 | (−) | NO_2 ⟶ CH_2CH_3 | O | O_2N | NO_2 | NO_2 $+ \; CH_3O^-$

Recently the structure of the Meisenheimer complex has been analyzed by X-ray crystallography,[4] as shown in Fig. 3.1, suggesting that methoxide as a donor complexes with 2,4,6-trinitroanisole as an acceptor to form a covalently bonded σ complex ($D^+–A^-$) by sp^3 hydrization at C-1 of trinitroanisole through charge-transfer complexing. The two alkoxy groups bind equivalently with C-1.

The rate constants for substitution have been measured[5] between 1,3,5-trinitrobenzene and various nucleophilic reagents and are in the following decreasing order:

$$C_2H_5O^- > CH_3O^- > OH^-.$$

The kinetics of formation of the π complex in the substitution reaction have been studied by stopped-flow spectrometry.[6] The formation of π complex (TNB·2Am) is reversible in the system between 1,3,5-trinitrobenzene(TNB) and dimethylaniline(Am). The rates of π-complex formation follow first-order kinetics with respect to the concentration of trinitroben-

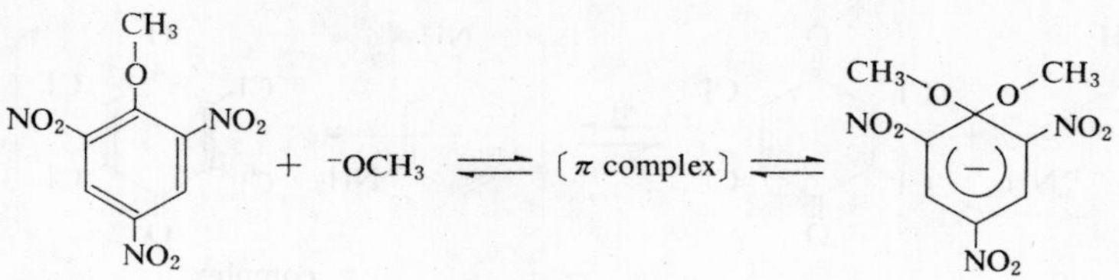

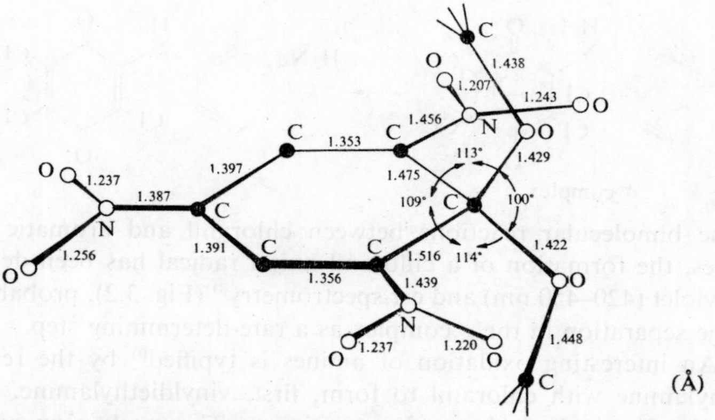

$$\text{2,4,6-}(NO_2)_3C_6H_2OCH_3 + {}^{-}OCH_3 \rightleftharpoons [\pi \text{ complex}] \rightleftharpoons [(CH_3O)_2C_6H_2(NO_2)_3]^{-}$$

Fig. 3.1 Structure of the Meisenheimer complex between methoxide ion and 2,4,6-trinitroanisole.
(Source: R. Destro, C. M. Gramaccioli and M. Scinonneta, *Acta Cryst.*, **24**, 1378 (1968). Reproduced by kind permission of the International Union of Crystallography, England.)

zene, and with respect to dimethylaniline. The half-life of the π complex has been measured as less than 0.05–1 sec at temperatures ranging from $-30°$ to $-85°C$ by rapid-scanning ultraviolet spectrometry.

Nagakura *et al.*[7] investigated in detail the mechanism of organic reactions through π and σ complex formation as reaction intermediates.

When a solution of *m*-phenylenediamine as a donor is mixed with chloranil at low temperatures, rapid spectral changes are observed due to the formation of 1:1 and 1:2 adducts as substitution products. A new absorption peak at 680 nm appeared after mixing both solutions at $-70°C$. This can be assigned to the formation of a π complex (outer complex). This absorption peak disappeared instantaneously on rewarming to room temperature, and other new absorption peaks were observed at 500 and 380 nm; that at 500 nm was assigned to the formation of a Meisenheimer-type σ complex (inner complex), and that at 380 nm may represent the formation of 1:1 adduct, as shown in the following scheme:

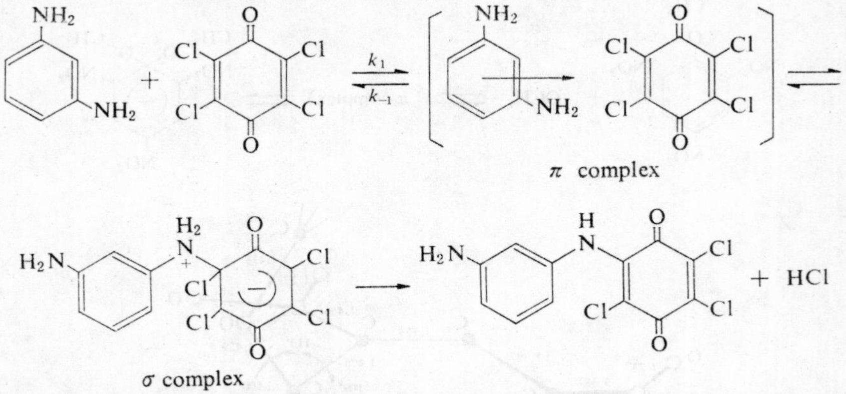

k_1, k_{-1}

π complex

σ complex

+ HCl

In the bimolecular reactions between chloranil and aromatic or alkyl amines, the formation of a chloranil anion radical has been detected by ultraviolet (420–450 nm) and esr spectrometry[8] (Fig. 3.2), probably due to charge separation in the π complex as a rate-determining step.

An interesting oxidation of amines is typified[9] by the reaction of triethylamine with chloranil to form, first, vinyldiethylamine, and then diethylaminovinyl-trichloro-*p*-benzoquinone. The mechanism and course of the reaction are illustrated in Fig. 3.3. The reaction is very sensitive to

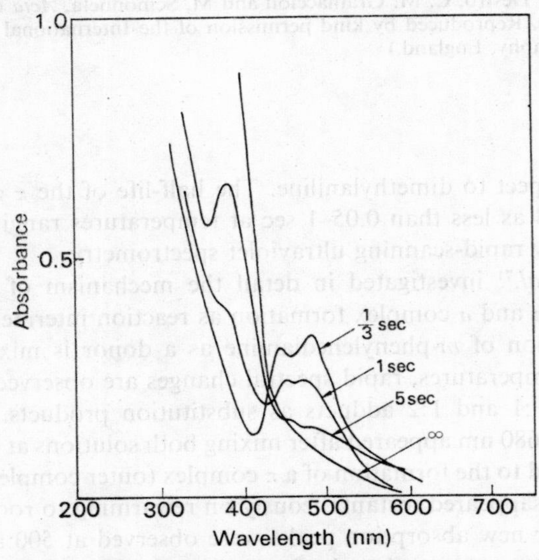

Fig. 3.2 Spectral changes in the system between *n*-butylamine and chloranil. The new absorption peak at *ca.* 420–450 nm corresponds to the formation of chloranil anion radical; the new absorption peak at 360 nm is assigned to the 1:1 adduct with 2,5-dibutylamino-3,6-dichloro-*p*-benzoquinone.

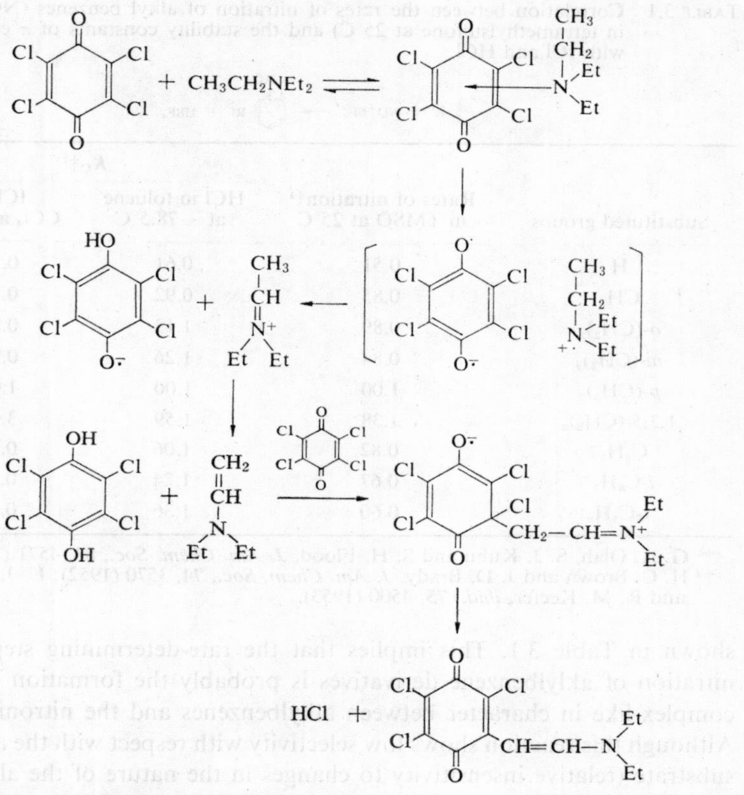

Fig. 3.3 Reaction of triethylamine with tetrachloro-*p*-benzoquinone.

steric factors: tetraiodo-*p*-benzoquinone does not react with triethylamine although the corresponding tetrabromoquinone reacts readily. While a hydride-transfer mechanism might be sensitive to the steric requirements of the quinone, even though the critical approach should be at the oxygen atom and in a direction away from the ring, it is not surprising that an electron-transfer reaction would be sensitive to the separation of the partners in the resulting ion pair. One can estimate that the Coulombic interaction in the ion pair would be reduced by 3 kcal/mole for each increase of 0.1 Å in average separation. The reaction of chloranil with other amines was also selective in a way that suggested a steric requirement for the overall reaction (i.e. dehydrogenation).

There is a good correlation[10] between the relative rates of nitration of substituted alkylbenzenes with NO_2^+ BF_4^- and the stability constants of the π complexes (K_c) of the corresponding alkylbenzenes with ICl, as

TABLE 3.1 Correlation between the rates of nitration of alkyl benzenes ($NO_2^+ BF_4^-$ in tetramethylsulfone at 25°C) and the stability constants of π complexes with ICl and HCl

$$C_6H_5R' + NO_2^+BF_4^- \longrightarrow (NO_2)C_6H_4R' + HBF_4$$

Substituted groups	Rates of nitration†[1] in TMSO at 25°C	K_C†[2] HCl in toluene at −78.5°C	K_C†[2] ICl in CCl_4 at 25°C
H	0.51	0.61	0.36
CH_3	0.85	0.92	0.57
o-$(CH_3)_2$	0.89	1.13	0.82
m-$(CH_3)_2$	0.84	1.26	0.92
p-$(CH_3)_2$	1.00	1.00	1.00
1,3,5-$(CH_3)_3$	1.38	1.59	3.04
$C_2H_5^-$	0.82	1.06	0.58
i-$C_3H_7^-$	0.67	1.24	0.58
t-C_4H_9	0.60	1.36	0.58

†[1] G. A. Olah, S. J. Kuhn and S. H. Flood, *J. Am. Chem. Soc.*, **83**, 4571 (1961).
†[2] H. C. Brown and J. D. Brady, *J. Am. Chem. Soc.*, **74**, 3570 (1952); L. J. Andrews and R. M. Keefer, *ibid.*, **75**, 4500 (1953).

shown in Table 3.1. This implies that the rate-determining step in the nitration of aklylbenzene derivatives is probably the formation of the π complex-like in character between alkylbenzenes and the nitronium ion. Although this reaction shows low selectivity with respect with the aromatic substrate (relative insensitivity to changes in the nature of the alkyl substituents), the monosubstitution product obtained by nitration of toluene with $NO_2^+ BF_4^-$ contains very little of the *meta* isomer (65.4% *ortho*, 2.8% *meta* and 31.8% *para*). Olah, Kuhn and Flood suggested[10)] that a π complex, in which the NO_2^+ is preferentially oriented to the most electron-rich spots on the aromatic nucleus, is formed as an intermediate in the slow step of the reaction. In contrast, the relative rates of halogenation of alkylbenzene derivatives with Cl_2 or ICl depend mainly upon the methylated benzenes used as donors, and correlate well with the stability constants of the σ complexes with $HF–BF_3$ rather than those of π complexes with HCl.[11)] The alkylated benzenes form a protonated σ complex with strong Brønsted acids such as $HF–BF_3$ in the following reaction:

$$ArR + HX \rightleftharpoons RArH^+X^- \quad \text{e.g.} \left[\text{(R,H)}C_6H_5^{(+)} \right] X^- \quad X{=}BF_4^-$$

In the latter case, protonated σ complex formation is estimated to be the rate-determining step for the halogenation of aromatic compounds.

Nagakura *et al.*[12] have proposed a working hypothesis to determine the rate and orientation of bimolecular nucleophilic (or electrophilic) substitution under consideration of the extent and direction of charge transfer between the donor and acceptor reagents. The formation of π and/or σ complexes as intermediates in the substitution reactions correlates with the extent of charge transfer, which is controlled by the relative energy separation between the highest occupied (*H. O.*) orbital levels of donor reagents and the lowest vacant (*L. V.*) orbital levels of acceptor reagents. The electrophilic reaction occurs thermally if the *L. V.* orbitals of acceptor reagents such as NO_2^+, Cl^+, etc. exist below the *H. O.* orbitals of donor reagents such as benzene, alkylbenzene, etc. The relative energy levels of *H. O.* and *L. V.* orbitals of some reagents can be estimated simply from physical values such as I_p of donors and EA (electron affinity) of acceptors, as presented schemetically in Fig. 3.4.

It is noteworthy that the extent and direction of charge transfer also depend upon the symmetries of the *H. O.* and *L. V.* orbitals of the reagents and the polarization energies. Nucleophilic reagents such as NO_2^+ and Br^+ thus initiate the substitution reaction with benzene, but Ag^+ and I_2

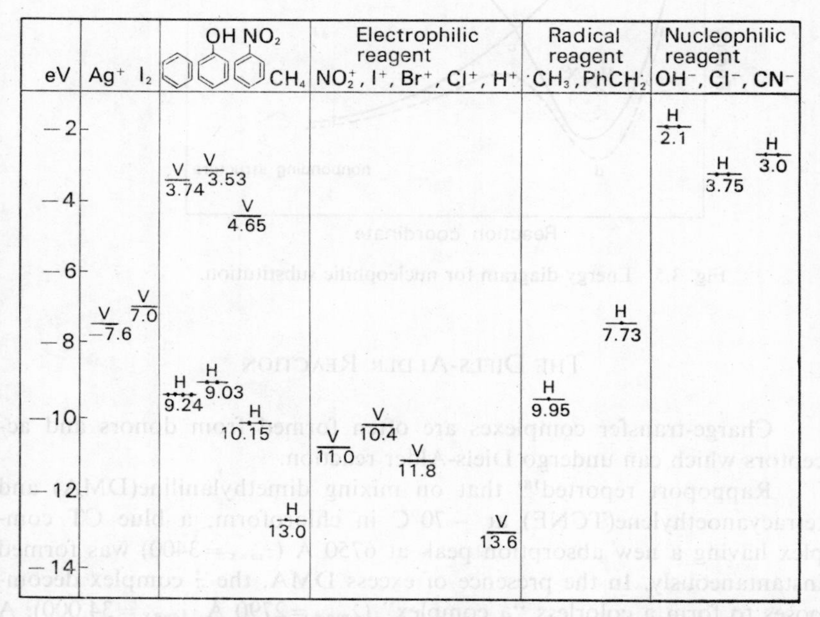

Fig. 3.4 Relative energy diagram for the *H O* and *L V* orbital energy levels of nucleophilic and electrophilic reagents.

(although they are also electrophilic) do not react, as would be expected from the relative energy levels shown in Fig. 3.4.

The course of the aromatic substitution reaction is shown schematically by the energy diagram (Fig. 3.5), where the solid lines represent potential curves for the nonbonding and dative structures between benzene and an electrophilic reagent X. Resonance stabilization among the nonbonding and dative structures changes the potential curve to that shown by the broken line. In advance of the charge-transfer interaction between benzene and X^+ the contribution of the dative structure increases to near 50% at the point (c), through which the relatively stable σ complex is formed (d).

In the transition state, the electrophilic reagent X^+ and reactant (such as benzene) act like radical and radical cation species, respectively. The orientation of substitution might be estimated from the electron density distribution of the radical cations of benzene derivatives formed as intermediates.

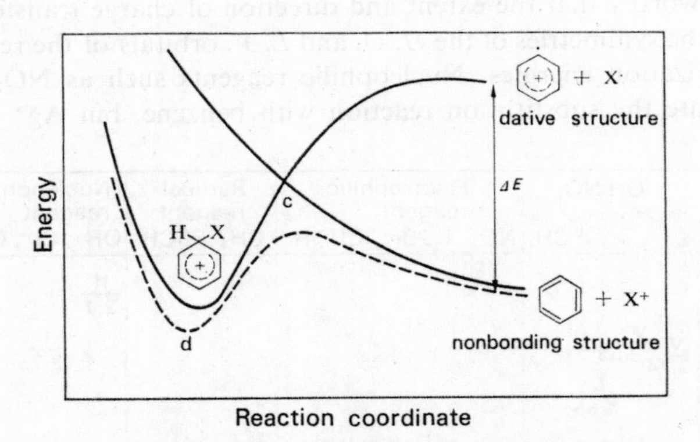

Fig. 3.5 Energy diagram for nucleophilic substitution.

The Diels-Alder Reaction

Charge-transfer complexes are often formed from donors and acceptors which can undergo Diels-Alder reaction.

Rappoport reported[13] that on mixing dimethylaniline(DMA) and tetracyanoethylene(TCNE) at −70°C in chloroform, a blue CT complex having a new absorption peak at 6750 Å ($\varepsilon_{max}=3400$) was formed instantaneously. In the presence of excess DMA, the π complex decomposes to form a colorless "σ complex" ($\lambda_{max}=2790$ Å, $\varepsilon_{max}=34{,}000$). A further slow reaction of the "σ complex" with DMA gave the final product,

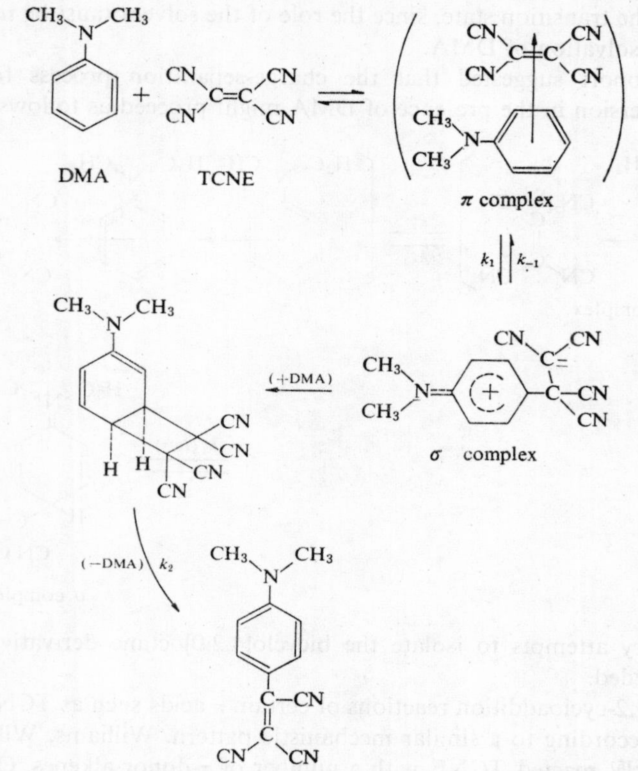

Fig. 3.6 Proposed mechanism for the reaction of DMA and TCNE.

N,N-dimethyl-4-tricyanovinylaniline; this part of the reaction was irreversible. The general scheme is shown in Fig. 3.6, where a bicyclo[4.2.0]octane derivative has been postulated as an intermediate produced from the "σ complex" ion-radical pair to yield the final product, *N,N*-dimethyl-4-tricyanovinylaniline.

Both the rate constants k_1 (π–σ conversion) and k_2 (formation of the final product) were found[14)] to be very sensitive to solvent, although only a limited range of solvents was actually used. Over the range of solvent polarity from 80% CCl_4–$CHCl_3$ to 80% $ClCH_2CH_2Cl$–$CHCl_3$, k_1 increased by a factor of about 150, while k_2 increased by a factor of 50. It should also be mentioned that these rates were measured in the presence of a large excess of DMA. If DMA is functioning in the present case as a π donor, the first-order dependence of the π–σ conversion (k_1) on [DMA] could imply that the large solvent sensitivity is indicative of a large charge separation in the transition state, since the role of the solvent must be reduced by such π solvation of DMA.

Rappoport suggested that the charge-separation process for the $\pi \rightarrow \sigma$ conversion in the presence of DMA might proceed as follows:

$$\left[\text{DMA} \rightarrow (CN)_2C{=}C(CN)_2 \right]_{\pi\ \text{complex}} \underset{k'_{-1}}{\overset{k'_1}{\rightleftharpoons}} \left[\text{DMA} \rightarrow \overset{+}{\text{DMA}} \rightarrow (CN)_2\dot{C}{-}\overset{-}{C}(CN)_2 \right] \underset{}{\overset{k''_1(\text{fast})}{\rightleftharpoons}} \sigma\ \text{complex}$$

π complex

σ complex

Preliminary attempts to isolate the bicyclo[4.2.0]octane derivative have not succeeded.

The 1,2-cycloaddition reactions of certain π acids such as TCNE may proceed according to a similar mechanistic pattern. Williams, Wiley and McKusick[15] reacted TCNE with a number of π-donor alkenes. On mixing, the reactants formed colored solutions probably due to charge-transfer complex (π-complex) formation. The color faded fairly quickly between 0 and 30°C, and the cycloaddition product could be isolated from the resulting solution in high yield. Some typical examples are shown below.

$$CH_3{-}O{-}CH{=}CH_2 + (CN)_2C{=}C(CN)_2 \overset{\text{THF}}{\rightleftharpoons} (\text{red-orange } \pi \text{ complex}) \longrightarrow \text{1-methoxy-2,2,3,3-tetracyanocyclobutane}$$

$$\text{dihydropyran} + (CN)_2C{=}C(CN)_2 \overset{\text{THF}}{\rightleftharpoons} (\text{red-orange } \pi \text{ complex}) \longrightarrow \text{tetracyano-2-oxabicyclo[4.2.0]octane}$$

$$CH_3\text{-}O\text{-}C_6H_4\text{-}CH{=}CH_2 + (CN)_2C{=}C(CN)_2 \underset{}{\overset{THF}{\rightleftharpoons}} \text{(blue } \pi \text{ complex)}$$

$$\longrightarrow CH_3\text{-}O\text{-}C_6H_4\text{-}\text{(cyclobutane with CN, CN, CN, CN)}$$

The rate of 1,2-cycloaddition is very much dependent upon the polarity of the solvent. A qualitative measure of the reaction rate was obtained[16] by observing the time required for the disappearance of the blue color of the *p*-methoxystyrene–TCNE complex in various solvents. (There is no assurance, of course, that the extent of complexing was the same in each solvent or that the absorption coefficient of the complex was the same in all solvents.)

The times were as follows: formic acid, 12 sec; nitromethane, 60 sec; nitrobenzene, 300 sec; ethanol, 8000 sec; diethyl ether, 20,000 sec; ethylacetate, 80,000 sec; toluene 600,000 sec; cyclohexane and carbon tetrachloride, more than 2,600,000 sec. The sensitivity to solvent polarity is sufficiently great to require a large charge separation in the transition state, arising from an initial state in which there is only a modest separation of charge. An electron-transfer process followed by the collapse of the resulting diradical may be a reasonable explanation for the reaction scheme, as shown in Fig. 3.7.

An important qualification to the π–σ conversion in 1,2-cycloaddition has been given by Proskow, Simmons and Cairns[17] in the reaction system between *cis*- and *trans*-1,2-bis(trifluoromethyl)-1,2-dicyanoethylene and alkene donors. Vinyl-*t*-butylsulfide reacts with each geometric isomer to give a cyclobutane derivative with absolute stereospecificity. The rates vary by as much as 10^{15} over the range of solvent polarity in the cycloaddition reaction. Accordingly, the proposed ion-pair diradical intermediates cannot have very much rotational freedom with respect to one another, or collapse to the final product is achieved in such a short time that rotation does not occur (Fig. 3.8)

In the 1,2-cycloaddition between TCNE and anthracene a green CT complex has been isolated, which yields a dibenzobicyclo[2.2.2]octane derivatives as the final product. Brown *et al.*[18] have succeeded in analyzing the complete crystal structure of the adduct product. As shown in Fig. 3.9, TCNE can overlap its double bond onto the 9,10-electron-rich position of anthracene, implying that 1,2-cycloaddition proceeds in keeping with the

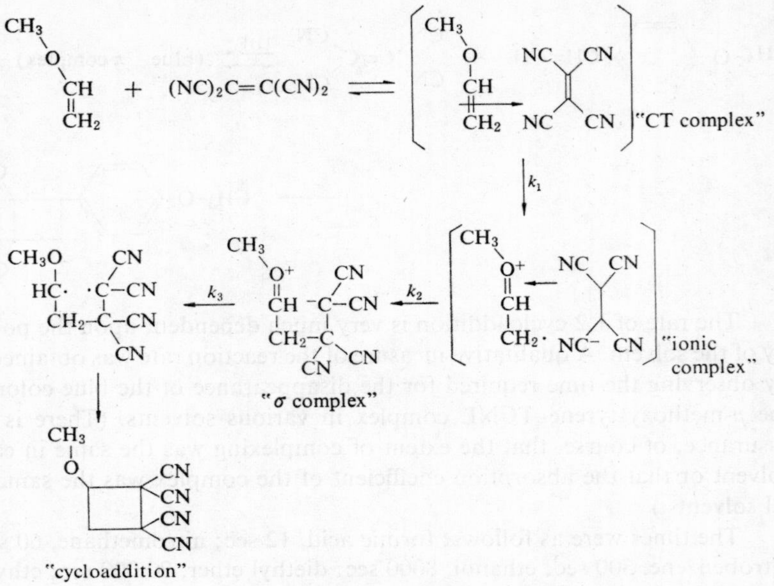

CH_3–O–CH=CH_2 + $(NC)_2C{=}C(CN)_2$ ⇌ [CH_3–O–CH=CH_2 → $(NC)_2C{=}C(CN)_2$] "CT complex"

k_1 ↓

[CH_3–O^+=CH–CH_2· NC^- CN, NC CN] "ionic complex"

k_2 ← CH_3–O^+=CH–C(CN)_2–CH_2–C(CN)_2 "σ complex"

k_3 ← CH_3O–HC·–CH_2–C(CN)_2–·C(CN)_2

↓

CH_3–O–(cyclobutane)–(CN)_4 "cycloaddition"

Fig. 3.7 Proposed mechanism for 1,2-cycloaddition of donor alkenes to acceptor alkenes.

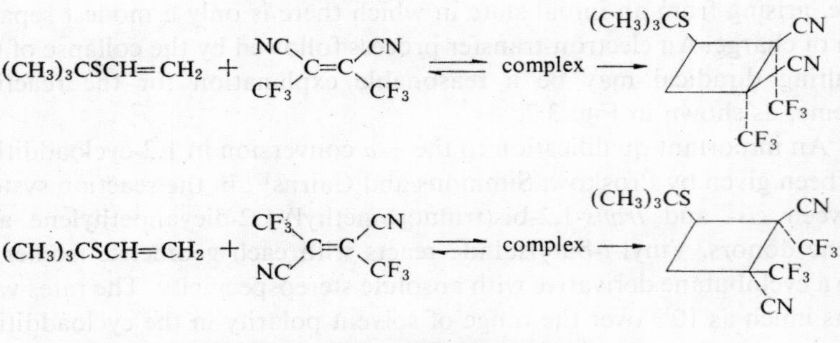

$(CH_3)_3CSCH{=}CH_2$ + NC(CF$_3$)C=C(CN)CF$_3$ ⇌ complex → cyclobutane with $(CH_3)_3CS$, CN, CN, CF_3, CF_3

$(CH_3)_3CSCH{=}CH_2$ + CF$_3$(NC)C=C(CN)CF$_3$ ⇌ complex → cyclobutane with $(CH_3)_3CS$, CN, CF_3, CF_3, CN

Fig. 3.8 Cycloaddition via complex formation.

stereospecific conformation of the CT complex between the donor and acceptor components. Thus, the initial complex (π complex) may play a role in determining the relative orientation of the donor and acceptor in the cycloaddition.

Martin and Hill have stated[19] that 1,2-cycloaddition between cyclopentadiene and maleic anhydride involves selective formation of the *endo* product, although the *exo* would be expected in a thermal reaction according to the Woodward-Hoffman rules. The formation of the *endo* cyclo-

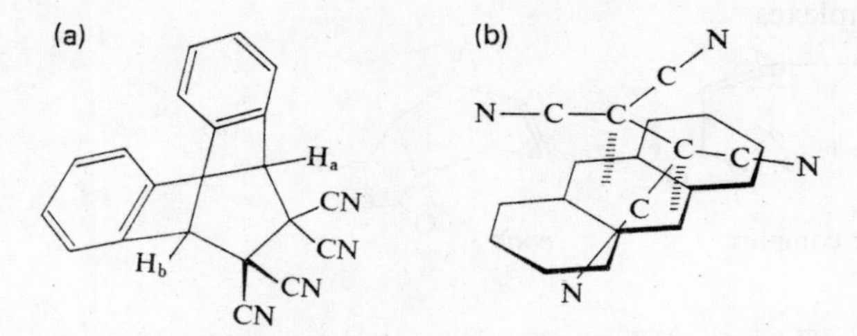

Fig. 3.9 (a) Structure of the addition compound; (b) proposed configuration of the CT complex between anthracene and TCNE.

addition product would be due to stabilization of the *endo* structure by CT complexing between cyclopentadiene and maleic anhydride in the transition state, as shown in Fig. 3.10.

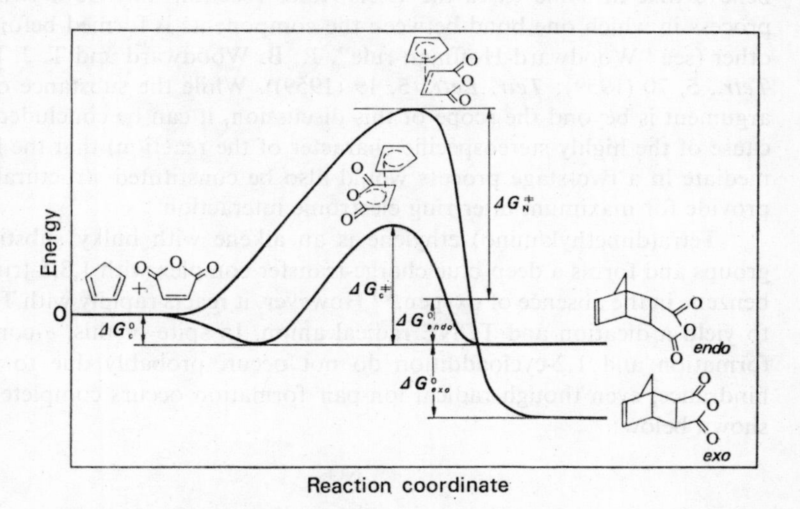

Fig. 3.10 Energy diagram for cycloaddition.

The stereochemistry of the products is generally correctly predicted by assuming that combination occurs preferentially when the two reactants are so oriented as to provide for maximum overlap of their unsaturated centers. In this connection the fact that the *endo* adduct of cyclopentadiene and maleic anhydride is formed much more rapidly than the *exo* isomer is often cited (see diagrams (a) and (b)). Since this same orientation should provide for maximum stabilization of a donor-acceptor complex of the diene and dienophile, it is reasonable to consider seriously the possibility that the complexes

π complex — endo (a)

exo (b)

are true intermediates in the Diels-Alder reactions. There is reason to believe that in some cases the Diels-Alder reaction may be a two-step process in which one bond between the components is formed before the other (see "Woodward-Hoffman rule", R. B. Woodward and T. J. Katz, *Tetr.*, **5**, 70 (1959); *Tetr. Lett.*, **5**, 19 (1959)). While the substance of the argument is beyond the scope of this discussion, it can be concluded (because of the highly stereospecific character of the reaction) that the intermediate in a two-stage process would also be constituted structurally to provide for maximum inter-ring electronic interaction.

Tetra(dimethylamino) ethylene is an alkene with bulky substituted groups and forms a deep-blue charge-transfer complex with 1,3,5-trinitrobenzene in the absence of oxygen.[20] However, it reacts rapidly with TCNE to yield a dication and TCNE radical anion. In spite of this, σ-complex formation and 1,2-cycloaddition do not occur, probably due to steric hindrance, even though radical ion-pair formation occurs completely, as shown below:

$$((CH_3)_2N)_2C{=}C(N(CH_3)_2)_2 + C_6H_3(NO_2)_3 \rightleftharpoons \pi \text{ complex}$$

$$((CH_3)_2N)_2C{=}C(N(CH_3)_2)_2 + 2\,(NC)_2C{=}C(CN)_2 \longrightarrow$$

$$((CH_3)_2N)_2\overset{+}{C}{-}\overset{+}{C}(N(CH_3)_2)_2 + 2\,(NC)_2\dot{C}{-}\overset{-}{C}(CN)_2$$

Ionic complex

3.2 Reactions Favored by the Characteristic Geometry of EDA Complexes

There are many reactions in which a particular pathway is favored because of complex formation between the reactants. Stereospecific hydrolyses by enzymes can be explained in this way without the necessity for postulating any single type of force or interaction as being responsible for the special structure of the complex. In some instances, however, it seems reasonable or likely that a complex which exhibits charge-transfer light absorption (or a closely related complex) is an actual intermediate in the formation of certain products.

Although the degree of charge separation in the ground states of most charge-transfer complexes is small, the charge-transfer portion of the interaction which produces a complex is not negligible. The charge-transfer interaction plays a major role in ensuring favorable geometry in the donor and acceptor components in relation to a specific reaction pathway.

It would be rather difficult to draw a clear distinction between the donor-acceptor interaction presumed to operate in the transition state described above and those cases in which a metastable carbonium ion intermediate is formed by π participation. An interesting example of the latter case was found by Cram and Goldstein[21] who reported that [9]-paracyclophanyl-4-*p*-toluenesulfonate solvolyzes in acetic acid 1800 times as rapidly as cyclohexyl-*p*-toluenesulfonate, a rate factor almost as large as that observed for *cis*-5-cyclodecenyl-*p*-toluenesulfonate, for which the factor is 3770. The model shown for this hydrocarbon provides for minimum H–H and ring–H repulsion. The rate factors have been measured at 75°C for various [9]paracyclophanyl derivatives with different (2-, 3-, 4- and 5-) *p*-toluenesulfonate substitution.

TsO

AcOH (75°C) −TsO

[9]paracyclophanyl-5-*p*-toluenesulfonate

The relative rates of solvolysis in relation to cyclohexyl-*p*-toluenesulfonate are 2, 9, 1800 and 170 times for the 2-, 3-, 4- and 5-*p*-toluenesulfonates of paracyclophanyl derivatives, respectively. Of these, 4-*p*-toluenesulfonate was solvolyzed most rapidly, probably due to stabilization by the partial positive charge generated in solvolysis with the benzene ring (as a donor group) in a favorable conformation near the cation center.

Apparently the geometric requirements for the stabilization of a transition state are more stringent for the chemical reaction than those for the occurrence of charge-transfer absorption, as suggested by the work of White on *cis*- and *trans*-1-(4-aminophenyl)-2-(4-nitrophenyl)cyclopentane (see section 2.1).

Kosower[22)] has studied the possible role of charge-transfer interaction in the reduction of nicotinamide derivatives with $Na_2S_2O_4$. The important coenzyme, nicotinamide adenine dinucleotide (NAD) is reduced by sodium dithionite, $Na_2S_2O_4$, to dihydrocoenzyme, NADH, which is formed as a result of enzymatic reduction (with ethanol and alcohol dehydrogenase, for example) involving stereospecific hydrogen transfer. Reduction of NAD by other means, e.g. electrochemical[23)] or with sodium borohydride,[24)] leads to mixtures with low or zero coenzymatic activity. Such reductions are illustrated[25)] by the behavior of the 1-propyl-3-carbamidopyridinium ion, as shown in Fig. 3.11.

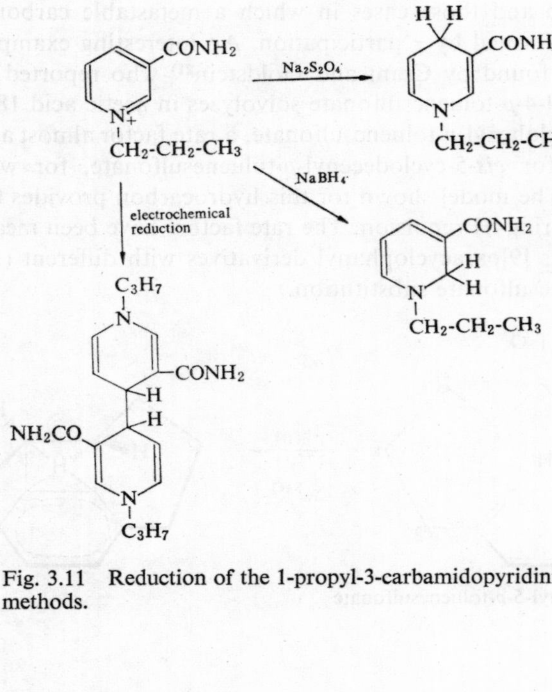

Fig. 3.11 Reduction of the 1-propyl-3-carbamidopyridinium ion by various methods.

In moderately basic solution, NAD (and other pyridinium ions) reacts with dithionite ions to yield a deep-yellow or orange solution from which NADH may be obtained by adjustment of the pH toward neutrality.[26] It was shown by Yarmolinsky and Colowick[27] that the yellow intermediate is formed fairly rapidly (but not instantaneously) according to the equation

$$NAD^+ + S_2O_4^{2-} \longrightarrow NAD{\cdot}SO_2^- + SO_2\,.$$

The intermediate was postulated to be a 1,4-dihydropyridine derivative (**1**) on the basis of its spectrum. A careful analysis of the spectrum of $NAD{\cdot}SO_2^-$, given elsewhere, suggested that the absorbing species was a charge-transfer complex. The complex proposed as the intermediate in the dithionite reduction of 1-alkylcarbamino-pyridinium ions (i.e. NAD) by Kosower and Bauer[28] is shown below as (**1**), while the isomeric complex derived from a 1-alkyl-4-carbamidopyridinium ion is shown as (**2**). According to the criterion suggested in section 2.2, the charge-transfer bands of these two π complexes (with the same donor and two similar acceptors) should differ in transition energy, as shown in Table 3.2. The agreement between the transition-energy differences for the two sets of complexes, shown in Table 3.2, is excellent and supports the view that the intermediates in dithionite reduction are sulfonate–pyridinium ion CT complexes, which is converted to σ-complex (**3**) as the intermediate to produce 1,4-dihydropyridine by hydrolysis.

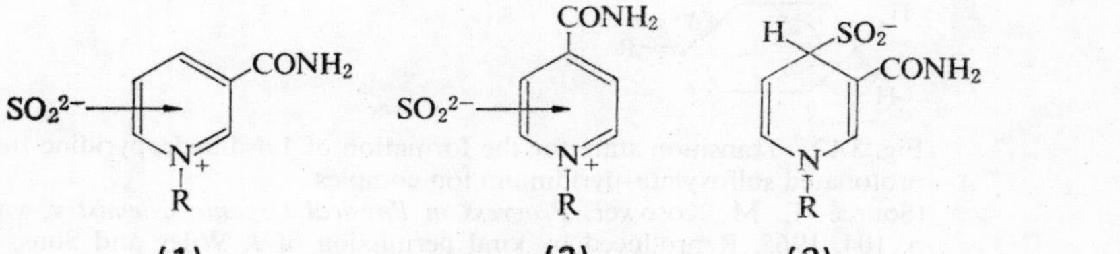

SO_2^{2-} → [pyridinium ring, 3-CONH$_2$, N$^+$–R] **(1)** SO_2^{2-} → [pyridinium ring, 4-CONH$_2$, N$^+$–R] **(2)** [ring, 4-H, 4-SO_2^-, 3-CONH$_2$, N–R] **(3)**

An attractive formulation[29] for the transition state leading to 1,4-dihydropyridine is shown in Fig. 3.12 and involves the transfer of hydrogen as a hydride ion from a protonated sulfoxylate to the pyridinium ring.

In a number of other cases, addition to the 4-position of the pyridinium ring is preferred, particularly with nucleophiles which one might suspect of being able to form complexes with the pyridinium ring in which a nucleophilic atom is near the 4-position and the anionic center can be located near the positively charged nitrogen. For example, the benzoylpyridinium ion reacts with acetophenone enolate anion to yield the 4-adduct. A reaction pathway has been proposed[30] based on the favorable geometry

TABLE 3.2 Charge-transfer bands for 1-alkyl-3- and 4-carbamidopyridine ion–sulfoxylate complexes

Pyridinium ion	λ_{max}	E_T (kcal/mole)	ΔE_T (kcal/mole)
1-Methyl-3-carbamido	3730	76.65	
1-Methyl-4-carbamido	4030	70.94	5.7 ± 0.3
1-Ethyl-3-carbamido	3720	76.85	
1-Ethyl-4-carbamido	4025	70.98	5.9 ± 0.3
NAD	3570	80.08	
iso-NAD†[1]	3870	73.88	6.2 ± 0.4
Value expected (difference for iodides)			6.7 ± 0.3†[2]

†[1] Isonicotinamide analog of NAD.
†[2] The difference would have been smaller in pure water.

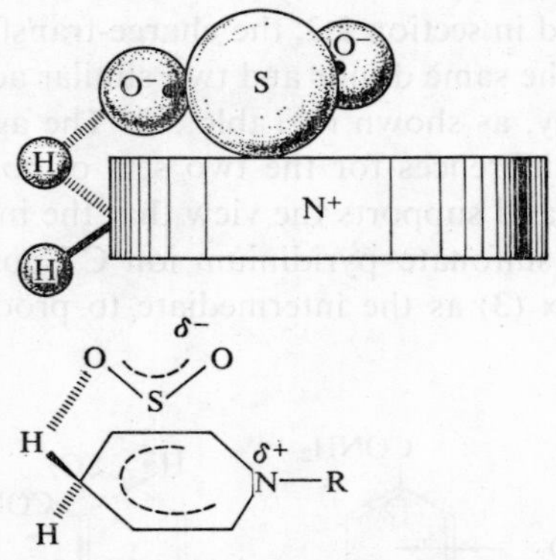

Fig. 3.12 Transition state for the formation of 1,4-dihydropyridine from a protonated sulfoxylate–pyridinium ion complex.
(Source: E. M. Kosower, *Progress in Physical Organic Chemistry*, vol. 3, p. 104, 1965. Reproduced by kind permission of J. Wiley and Sons, Inc., U.S.A.)

of the CT complex between the acetophenone enolate anion and benzoylpyridinium ion, yielding a σ complex, as shown in Fig. 3.13. The cyanide ion is the only exception to this generalization. Cyanide adds very easily to the pyridinium ring at the 4-position, but no charge-transfer absorption has been reported for such a combination.

Lansbury and Peterson[31)] have suggested that complex formation between the dihydropyridine ring of lithium tetrakis(*N*-dihydropyridyl) aluminate and an arylcarbonyl moiety is responsible for the rather specific reduction of 4-(4-benzoylphenyl)-2-butanone at the arylcarbonyl rather than the alkylcarbonyl group, for which sodium borohydride is effective.

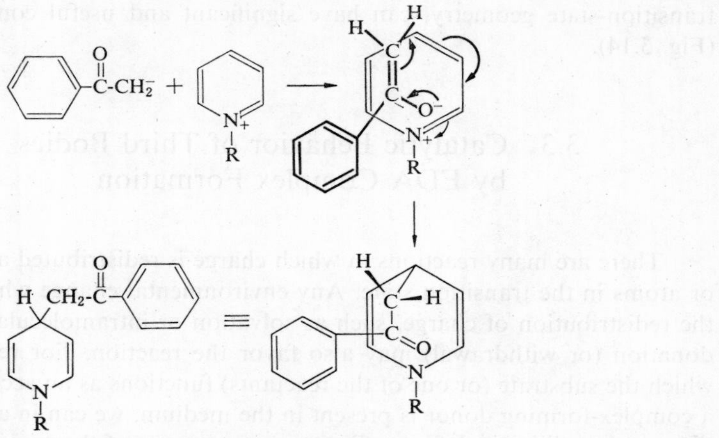

Fig. 3.13 Reaction of the benzoylpyridinium ion with acetophenone enolate anion.
(Source- E. M. Kosower, *Progress in Physical Organic Chemistry*, vol. 3, p. 106, 1965. Reproduced by kind permission of J. Wiley and Sons, Inc., U.S.A.)

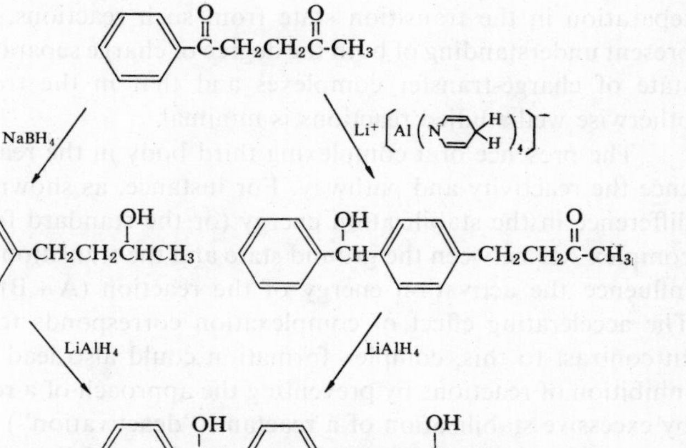

Fig. 3.14 Specific reduction of 4-(4-benzoylphenyl)-2-butane with lithium tetrakis(*N*-dihydropyridyl) aluminate.

Kosower and Sorensen[32] have found that 1,4-dihydropyridines can form charge-transfer complexes with arylcarbonyl derivatives such as benzophenone. The importance of this specific reduction, apart from its possible value as a synthetic tool, is that it illustrates (like the dithionate reduction) how relatively weak complex formation, presumably through its control of transition-state geometry, can have significant and useful consequences (Fig. 3.14).

3.3 Catalytic Behavior of Third Bodies by EDA Complex Formation

There are many reactions in which charge is redistributed at an atom or atoms in the transition state. Any environmental change which favors the redistribution of charge, such as solvation or intramolecular electron donation (or withdrawal) may also favor the reaction. For reactions in which the substrate (or one of the reactants) functions as an acceptor, and a complex-forming donor is present in the medium, we can in a few cases discern a small but a definite effect upon the course of the reaction. Reactions in which a donor (or acceptor) i.e. a third body, promotes the reactivity of a reactant acceptor (or donor) through charge-transfer complex formation are described as catalytic action by the third body. Eventually one might hope to extract information bearing upon the extent of charge separation in the transition state from such reactions, even though our present understanding of both the degree of charge separation in the ground state of charge-transfer complexes and that in the transition states of otherwise well-studied reactions is minimal.

The presence of a complexing third body in the reaction could influence the reactivity and pathway. For instance, as shown in Fig. 3.15, the difference in the stabilization energy (or the standard free energy) of the complex AD between the ground state and the transition state ($M^{\ddagger}$) would influence the activation energy of the reaction $(A+B)^{\ddagger}$ and $(AD+B)^{\ddagger}$. The accelerating effect of complexation corresponds to catalytic action. In contrast to this, complex formation could also lead to retardation or inhibition of reactions by preventing the approach of a reactant B to A, or by excessive stabilization of a reactant ("deactivation") by complexing.

In considering the reaction system of A and B in the presence of D, the following reaction scheme can be considered for the equibibrium state in thermodynamic terms.[33]

$$\begin{array}{ccccc} A+B+D & \overset{K_o^{\neq}}{\rightleftharpoons} & M_o^{\ddagger}+D & \longrightarrow & P+D \\ K \Big\uparrow\!\Big\downarrow \Delta G^{\circ} & & \Delta G^{\circ\neq} \Big\downarrow\!\Big\uparrow K_{\ddagger} & & \Big\downarrow\!\Big\uparrow K_p \\ AD+B & \underset{\Delta G_c^{\ddagger}}{\overset{K_c^{\neq}}{\rightleftharpoons}} & M_c^{\ddagger} & \longrightarrow & PD \end{array}$$

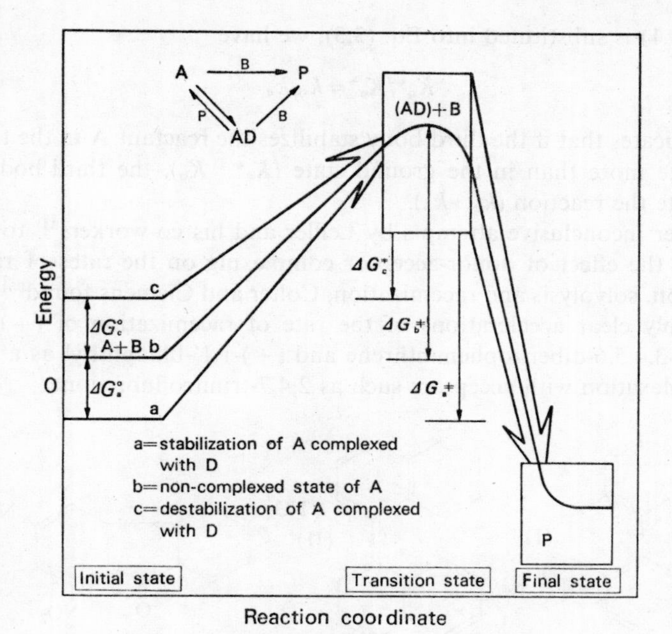

Fig. 3.15 Evolution of free energy in the course of a reaction by a complexing third body (D).
(After O. B. Nagy and J. B. Nagy, *Ind. Chim. Belg.*, 36, 940 (1971).

Here A and B are in equilibrium ($K_o^{\ne}$, $\Delta G_o^{\ne}$) with a noncomplexed transition state ($M_0^{\ne}$) to produce a product (P), which can also be produced via complexation ($K_c^{\ne}$, $\Delta G_c^{\ne}$, $M_c^{\ne}$). Supposing a transmission coefficient near one, the rate constants can be written as follows, where $K_0^{\ne}$, $K_c^{\ne}$, $\Delta G^{o}_{\ne}$, $\Delta G_c^{\ne}$, and K, ΔG^{o} are equilibrium constants and free energy changes of A and B in non-complexed and complexed transition states, and those for complexation between A and D.

$$k_o = kT/h\ K_o^{\ne} \quad \text{and} \quad k_c = kT/h\ K_c^{\ne} \tag{3.1}$$

The change of free energy is:

$$\Delta G_o^{\ne} = \Delta G^{o} + \Delta G_c^{\ne} - \Delta G^{o}_{\ne} \tag{3.2}$$

Thus

$$-RT \ln K_o^{\ne} = -RT \ln K - RT \ln K_c + RT \ln K_{\ne} \tag{3.3}$$

or $K_o^{\ne} = K \cdot K_c^{\ne}/K_{\ne}\,\{(M_o^{\ne})(D)/(A)(B)(C) = (AD)(B)/(A)(B)(C) \times M_c^{\ne}/(AD)(B) \times (M_o^{\ne})(D)/M_c^{\ne}\}$

If Eq. (3.1) is substituted into Eq. (3.3), we have

$$K_o^{\neq}/K_c^{\neq}=k_o/k_c \tag{3.4}$$

This indicates that if the third body stabilizes the reactant A in the transition state more than in the ground state ($K_c^{\neq}>K_o$), the third body will accelerate the reaction ($k_c>k_o$).

After inconclusive attempts by Leffler and his co-workers[34] to demonstrate the effect of donor-acceptor complexing on the rates of radical formation, solvolysis and racemization, Colter and Clemens found[35] some reasonably clear accelerations of the rate of racemization of (+)-9,10-dihydro-3,4,5,6-dibenzophenanthrene and (+)-1,1′-binaphthyl as a result of complexation with acceptors such as 2,4,7-trinitrofluorenone.

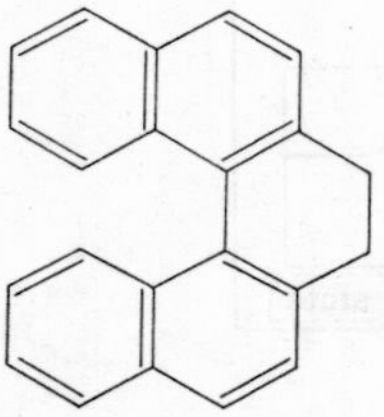

(I) (+)-9,10-dihydro-3,4,5,6-dibenzophenanthrene

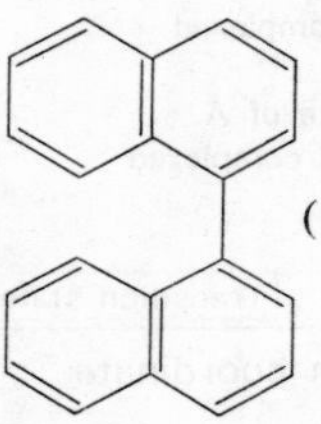

(II) (+)-1,1′-binaphthyl

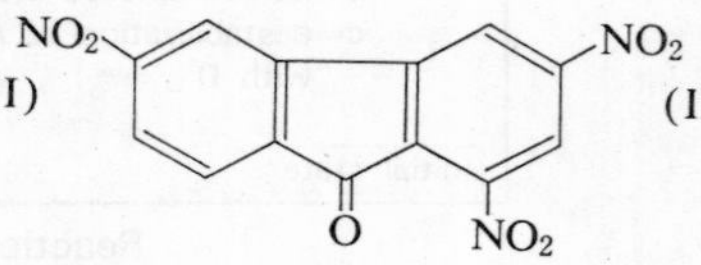

(III) 2,4,7-trinitrofluorenone

In nearly all instances, added acceptors produced rate enhancements. The largest effect observed was a 2.06-fold increase in the rate of racemization of I (dioxane, 33.68°C) produced by 0.227 M trinitrofluorenone (III). Catalytic effectiveness parallels acceptor strength for the two acceptors such as 1,3,5-trinitrobenzene and picryl chloride. These results are consistent with an explanation ascribing the rate enhancements entirely to charge-transfer complexing, only if the equilibrium constant for 1:1 complex formation with trinitrobenzene or picryl chloride at 33.7°C is approximately 0.5 l/mole^{-1} or smaller. Racemization of an optically active biphenyl, (I) and (II), is generally assumed to involve an approximately coplanar transition state; consequently we might expect the rate of racemization of a biphenyl donor to be increased by added acceptors.

Colter and his co-workers reported[36] that the acetolysis of 2,4,7-trinitro-9-fluorenyl-*p*-toluenesulfonate is accelerated by added aromatic molecules such as phenanthrene, hexamethylbenzene, acenaphthene and anthracene. These molecules are π donors and easily form π–π complexes with good acceptors. Acetic acid was used as a solvent. The reaction is illustrated in Fig. 3.16. The complexing constant (K_T) derived from the kinetic data using Eq. (3.5) is the same as that extrapolated from direct

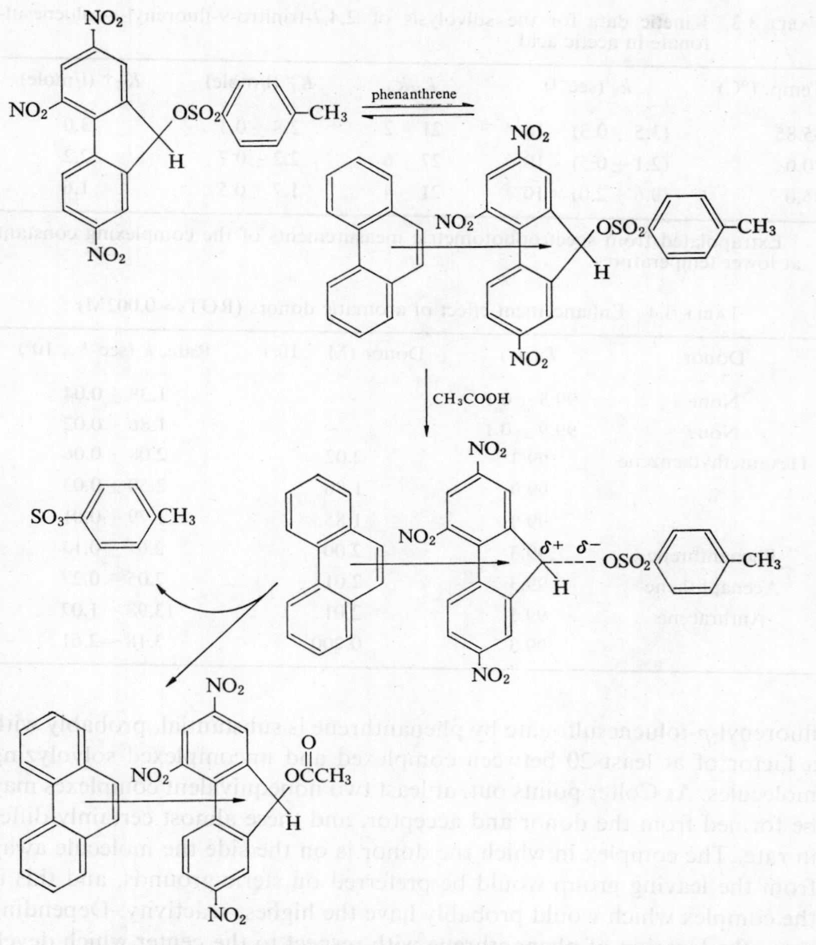

Fig. 3.16 Solvolysis of 2,4,7-trinitro-9-fluorenyl-*p*-toluenesulfonate in the presence of phenanthrene as a donor.

measurements of complex formation at lower temperature as shown in Table 3.3.

$$1/(k_{obs}-k_u)=1/(k_c-k_u)+1/K_T[D]_o(k_c-k_u) \qquad (3.5)$$

Here, k_{obs} = observed rate constant, k_c = rate constant for complexed toluenesulfonate, k_u = rate constant for uncomplexed toluenesulfonate, K_T = complexing constant, $(D)_o$ = donor concentration. It is clear from the data in Table 3.4 that the acceleration of the solvolysis of 2,4,7-trinitro-9-

TABLE 3.3 Kinetic data for the solvolysis of 2,4,7-trinitro-9-fluorenyl-*p*-toluenesulfonate in acetic acid

Temp. (°C)	k_c (sec^{-1})	k_c/k_u	K_T (l/mole)	K_T† (l/mole)
55.85	$(3.5 \pm 0.3) \times 10^{-6}$	21 ± 2	2.8 ± 0.3	3.0
70.0	$(2.1 \pm 0.5) \times 10^{-5}$	27 ± 6	2.2 ± 0.7	2.2
85.0	$(9.6 \pm 2.0) \times 10^{-5}$	21 ± 4	1.7 ± 0.5	1.6

† Extrapolated from spectrophotometric measurements of the complexing constant at lower temperatures.

TABLE 3.4 Enhancement effect of aromatic donors (ROTs=0.002M)

Donor	T (°C)	Donor (M $\times$ 10^2)	Rate, k (sec^{-1} $\times$ 10^5)
None	99.3 ± 0.1	—	1.38 ± 0.04
None	99.9 ± 0.1	—	1.86 ± 0.02
Hexamethylbenzene	99.3	2.02	2.08 ± 0.06
”	99.9	1.86	2.50 ± 0.03
”	99.9	1.85	1.79 ± 0.01
Phenanthrene	99.3	2.00	2.61 ± 0.13
Acenaphthene	99.3	2.01	7.05 ± 0.27
Anthracene	99.3	2.01	13.98 ± 1.07
”	99.3	0.200	$3.18 \sim 2.61$

fluorenyl-*p*-toluenesulfonate by phenanthrene is substantial, probably with a factor of at least 20 between complexed and uncomplexed solvolyzing molecules. As Colter points out, at least two nonequivalent complexes may be formed from the donor and acceptor, and these almost certainly differ in rate. The complex in which the donor is on the side the molecule away from the leaving group would be preferred on steric grounds, and this is the complex which would probably have the highest reactivity. Depending upon the location of phenanthrene with respect to the center which develops positive charge in the transition state, one would expect some degree of interference with the solvation necessary for the separation of charge to occur. If we assume that the second acetic acid molecule of the two which are probably most closely involved in solvating the transition state is displaced completely from its location near C-9 of the fluorenyl moiety, and that this second molecule is responsible for 25% of the solvation in the transition state, we can conclude that complexing is, in fact, responsible for a very large increase in rate over that which would be expected for the ester solvolyzing through a hypothetical transition state in which only one acetic acid molecule was close to C-9. Of course, the geometry of the complex may not require that the center of the 9,10-bond in phenanthrene

TABLE 3.5 Results of analysis of kinetic data for the dinitrofluorenyl-*p*-toluenesulfonates

Substrate	Temp. (°C)	$10^6 \times k_c$ (sec^{-1})	K (1/mole)	k_c/k_u
a	70.07	21 ± 5	2.2 ± 0.7	27 ± 6
b	50.17	77 ± 14	2.4 ± 0.5	12 ± 2
c	50.17	27 ± 3	2.4 ± 0.3	14 ± 2
d	50.17	6.4 ± 0.5	2.9 ± 0.3	11 ± 1
d	70.07	74 ± 13	1.7 ± 0.3	13 ± 2

TABLE 3.6 Structural effects on catalysis by charge-transfer complexing

Substrate†[1]	Temp. (°C)	Donor	[Donor] (M)†[2]	$10^6\ k$ (sec^{-1})†[3]	k_{obs}/k_u
e	30.1	none		9.26 ± 0.05	
e	30.1	phenanthrene	0.05	9.69 ± 0.05	1.05 ± 0.01
f	30.1	none		85.1 ± 0.8	
f	30.1	phenanthrene	0.05	87.1 ± 0.4	1.02 ± 0.01
g	50.17	none		88.3 ± 0.8	
g	50.17	phenanthrene	0.05	118 ± 1	1.34 ± 0.02
g	50.17	anthracene	0.02	172 ± 2	1.95 ± 0.04
g	50.17	hexaethylbenzene	0.02	89.0 ± 0.5	1.01 ± 0.01

†[1] Concn. of esters, 0.005 M for e and f; 0.002 M for g (for structure of e–g, see Table 3.5).
†[2] Measured at 25°C.
†[3] Listed with average deviations for six to ten measurements.

a: $R_1=R_3=NO_2$, $R_2=H$
b: $R_1=NO_2$, $R_2=R_3=H$
c: $R_1=R_3=H$, $R_2=NO_2$
d: $R_1=R_2=H$, $R_3=NO_2$
e: $R_1=R_2=R_3=H$

f: $R_1'=H$, $R_2'=NO_2$
g: $R_1'=NO_2$, $R_2'=H$

must lie over the 9-position of phenanthrene, but this is certainly a reasonable position if we compare the partial carbonium ion at C-9 to a silver ion and remember that the benzene–silver perchlorate complex has the silver ion located at an edge of the benzene ring.

The enhancement effects are presented in Table 3.4 for the solvolysis of 2,4,6-trinitro-9-fluorenyl-*p*-toluenesulfonate in the presence of various kinds of aromatic donors, implying that the enhancement effect increases almost proportionally with the concentration of donor and with the strength of electron-donating ability.[36,38)]

TABLE 3.7 Parameters of activation in the acetolysis of ROTs

$C_b(M^{-1})$	$\Delta H^{\neq}$ (kcal/mole)	$\Delta S^{\neq}$ (eu. mole^{-1})
0	25.7 ± 0.4	-11.7 ± 1.2
0.05	22.9 ± 0.3	-17.6 ± 0.8
0.08	22.4	-18.3

Poorer acceptors (2,4-dinitro- and 2,7-dinitrofluorenyl-*p*-toluenesulfonate) exhibit smaller, but still substantial, rate enhancements due to complexation with phenanthrene, with k_c/k_u values of about 12–13. (Table 3.5).[37]

In contrast, the rates of acetolysis of 3,3′- and 3,5-dinitrobenzhydryl-*p*-toluenesulfonates are essentially unchanged by 0.05 M phenanthrene, as shown in Table 3.6. The almost complete absence of any catalysis by 0.05 M phenanthrene in acetolysis of the two dinitrobenzhydryl tosylates is most likely due to the absence of appreciable complex formation. Both the geometry of the benzhydryl system (specifically, noncoplanarity of the benzene rings) and its lack of rigidity make complex formation unfavorable. From the results of this work,[37] it is clear that a negligible fraction of the rate enhancements in acetolysis of the dinitro- and trinitrofluorenyl tosylate is a result of chance encounters between substrate and donor.

Colter *et al.* have measured[36] the apparent activation enthalpy ($\Delta H^{\neq}$) and entropy ($\Delta S^{\neq}$) for the solvolysis of 2,4,7-trinitro-9-fluorenyl-*p*-toluenesulfonate in the presence of phenanthrene, as shown in Table 3.7. At extreme concentrations of donor phenanthrene, $\Delta H^{\neq}$ and $\Delta S^{\neq}$ have been determined to be ~21 kcal/mole and -18 to -19 eu, respectively, suggesting that the acceleration of the reaction due to the presence of donor molecules is mainly derived from the entropy change produced in the transition state by CT interaction. The *p*-toluenesulfonate acceptor exists in complexed form in the transition state with a larger apparent entropy than in the noncomplexed state.

Fig. 3.17 summarizes the energy profile of activation in the presence of a complexing third body (e.g. phenanthrene) for the acetolysis of 2,4,7-trinitrofluorenyl-9-toluenesulfonate. The value of -5 eu for $\Delta S_c^{\neq}$ shows that the reorganization energy of the solvent for complex formation at the transition state requires a smaller entropy than for free reactant ($\Delta S_0^{\neq} = -11.7$ eu). This also suggests that the difference of entropy change between the two systems $A+B \rightarrow (A\cdots B)^{\neq}$ and $AD+B \rightarrow (AD\cdots B)^{\neq}$ is due to the decrease of entropy activation accompanying CT complex formation in the presence of donor molecules. Thus the CT complex is an intermediate on the reaction pathway. Colter *et al.* have determined[37] the stability constants of the complex (AD) as $K_{\neq} = 55$ l/mole at 55.85°C in the transition state and $K = 3$ l/mole in the initial state.

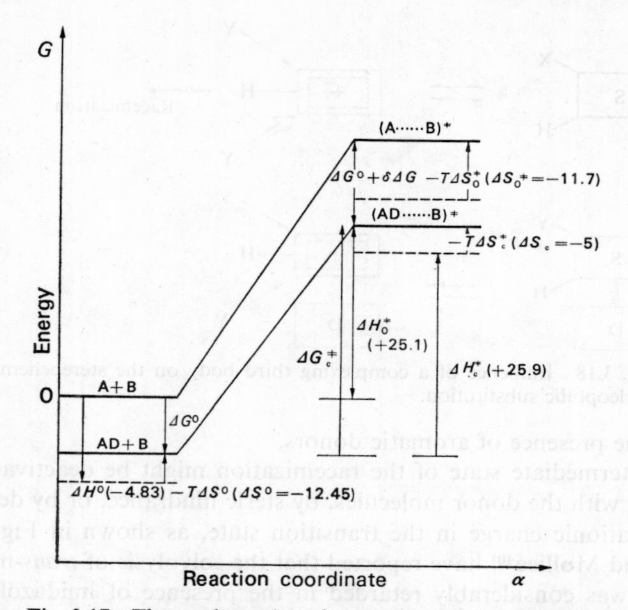

Fig. 3.17 Thermodynamics of activation in the presence of a complexing third body (acetolysis of 2,4,7-trinitrofluorenyl-9-*p*-toluenesulfonate). Units: ΔG, kcal/mole; ΔH, kcal/mole; ΔS, eu/mole.

Okamoto, Nitta and Shingu have studied[39] the acetolysis of (+)-camphane-10-sulfonate-2,4,7-trinitrofluorenyl-9 (at 11.64 ± 0.5°C in acetic acid). They observed that the reaction proceeded with almost complete retention of configuration. In the presence of aromatic donors such as hexamethylbenzene, phenanthrene and anthracene (D), the degree of optical rotation of the product decreased considerably. For example, the presence of hexamethylbenzene (7.11×10^{-2} mole/l) diminished the optical rotation to $-6.2°$ at -0.9°C. The kinetics of this reaction were similar to those proposed by Colter *et al.*[36] In Table 3.8 the different effects of acceleration on the racemization of the (+)-camphane-10 derivative are pre-

TABLE 3.8 Kinetic (T=116.4 ± 0.5°C) and spectroscopic (T=24.0°C) parameters of (+) camphane-10-toluenesulfonate-2,4,7-trinitro-9-fluorenyl in the presence of aromatic donors†

Donors	k_c ($\times 10^3$ sec^{-1})	k_c/k_o	K_T (l/mole)	K_{TUV} (l/mole)
Hexamethylbenzene	1.5	20	1.0	2.6
Phenanthrene	1.3	18	1.2	2.7
Anthracene	2.9	39	8 ± 4	4.1 ± 3

† $k_o = 7.5 \times 10^{-5}$ sec^{-1}; K_{TUV}=spectroscopic values for the stability constants of CT complexes of trinitrofluorenyl derivatives with aromatic donors.

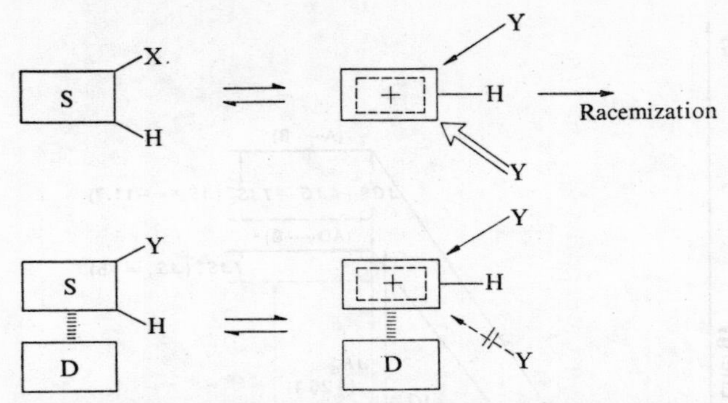

Fig. 3.18 Influence of a complexing third body on the stereochemistry of nucleophilic substitution.

sented in the presence of aromatic donors.

The intermediate state of the racemization might be deactivated by complexing with the donor molecules, by steric hindrance, or by delocalization of cationic charge in the transition state, as shown in Fig. 3.18. Connors and Mollica[40] have reported that the solvolysis of *trans*-methylcinnamate was considerably retarded in the presence of imidazole as a donor. They suggested that the retardation was probably due to the steric hindrance to the attack of OH^- on the carbonium cation center of cinnamate in the transition state of the solvolysis by CT-complex formation between cinnamate and imidazole. In fact, the formation of 1:1 CT complexes between imidazole (or imidazole derivatives) and *trans*-methylcinnamate has been reported by spectroscopy ($K=1.0\pm0.1$) (25°C, pH 7.2–8.3).

It has also been found[41] that other imidazole derivatives such as 2-methylimidazole, *N*-methylimidazole, benzylimidazole, purine, uracil, chloro-8-theophylline, theophylline and caffeine as electron donors, retard the solvolysis of *trans*-cinnamate in alkaline solution. In addition, they also form 1:1 CT complexes with methylcinnamate (Fig. 3.19).

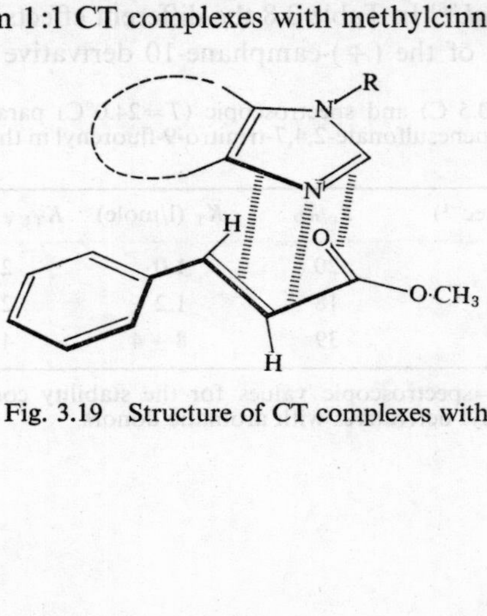

Fig. 3.19 Structure of CT complexes with methylcinnamate.

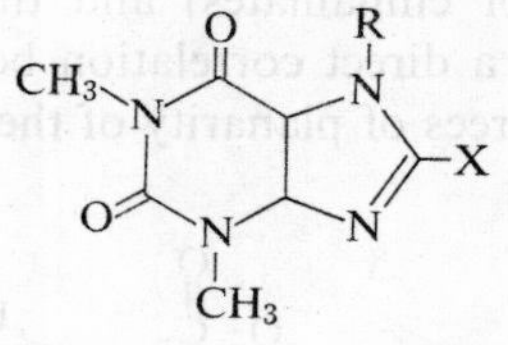

Xanthine derivatives:
Theophylline (R=X=H)
Halotheophylline (X=Cl, Br, I; R=H)
Dihydroxy-1,2,3-propyl-7-theophylline (X=H; R=CH_2–CH(OH)–CH_2(OH))
Caffeine (X=H; R=CH_3)
8-Methoxycaffeine (X=OCH_3; R=CH_3)

In all cases, the reaction occurs with retention of configuration due to some degree of complex formation.

Kramer and Connors[42] have also investigated the modification of the reactivity of several carboxylic acids (e.g. cinnamate) in the presence of complexants (e.g. xanthines). They analyzed the kinetic data on the retardation of the solvolysis of carboxylic acids due to 1:1 CT-complex formation. Alkaline hydrolysis of the carboxylic acids is inhibited by all the complexants; $q=0.24$ (theophylline), 0.18 (8-chlorotheophylline) and 0.66 (8-bromotheophylline) (pH 12.80 in 0.83% acetonitrile at 25.0°C, ionic strength 0.3). (Retardation coefficient $q=k_c/k^\circ$, where k_c and k° are the rate constants of alkaline hydrolysis in the presence and absence of complexants.) It is interesting to note that the inhibition of sulfonation of cinnamoyl double bonds is complete ($q=1$) in the presence of xanthine derivatives. The alkaline hydrolysis of *p*-nitrophenylbenzoate is almost completely inhibited in the presence of theophylline, as shown in Table 3.9, where the retardation ratios are shown for various nucleophilic reagents. The stability constant for the complex between *p*-nitrophenylbenzoate and theophylline has been measured as $K=120$ l/mole (pH 11.9, in H_2O at 25°C). Connors, Infeld and Kline[43] have examined a large number

TABLE 3.9 Hydrolysis of *p*-nitrophenylbenzoate in the presence of theophylline

Nucleophilic reagent	*K* (l/mole)	*q*
NH_2OH	20	0.61
NH_2NH_2	17	0.71
HOO^-	22	0.59
SO_3^{2-}	17	0.93
H_2O (OH^-)	12	1.0

of complexes between esters (mainly of cinnamates) and theophylline derivatives or their anions. They found a direct correlation between the free energy of complexation and the degrees of planarity of the theophylline derivatives.

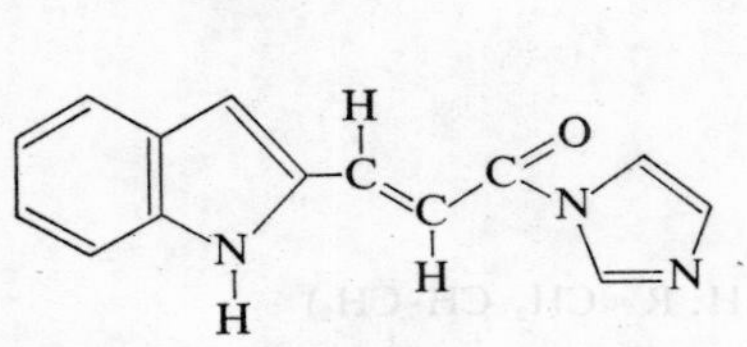

N-(indole-3-acryloyl)imidazole

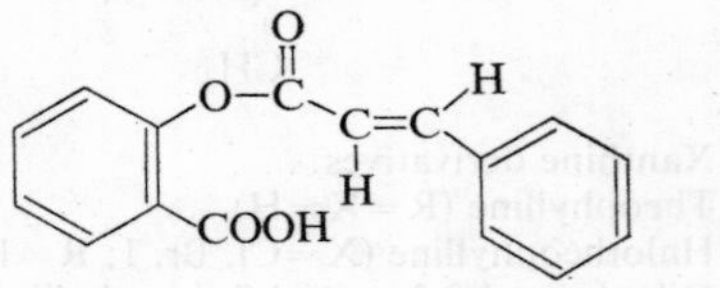

cinnamoylsalicylic acid

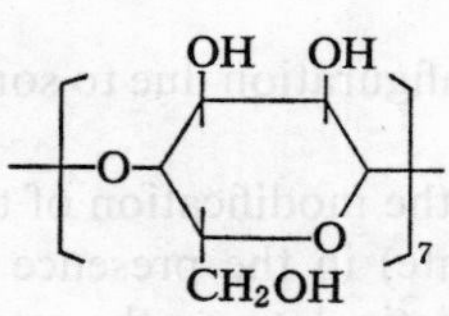

(α-D-glucopyranose)
β-cyclodextrin

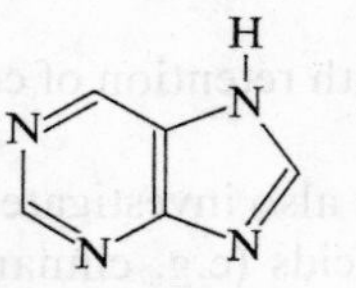

purine

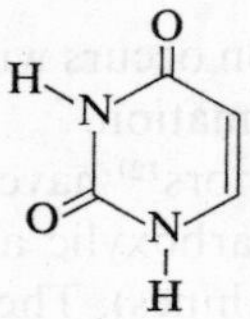

uracil

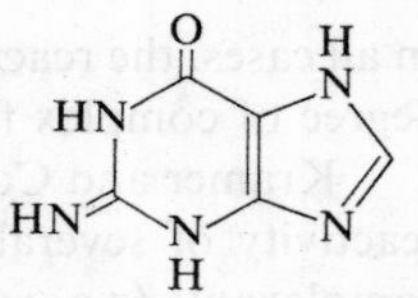

guanine

Su and Shafer[44] studied the inhibition of the alkaline hydrolysis of *N*-methylphthalimide by imidazole. They suggested that imidazole acts as a catalyst for the hydrolysis at low pH, but inhibits the reaction at pH = 9.7. They proposed that the imidazole forms a tetrahedral intermediate (σ complex) with imide, which inhibits the reaction for the following two reasons:

(1) The probability of attack on imide by OH^- might be reduced statistically due to steric hindrance in the complexation with imidazole.
(2) The *N*-CH_3 group might be affected considerably by complexation, so that the lone-pair electrons of the nitrogen atom cannot be readily delocalized over the carbonyl group.

The proposed mechanism is illustrated in Fig. 3.20. The inhibition might be also explained by the formation of a CT complex between imide and imidazole, preventing the attack of OH^-, but Shafer *et al*. have rejected that possibility. At lower pH values, imidazole reacts with imide as a nucleophilic reagent, like H_2O and OH^-, and enhances the rate of hydrolysis by attacking the carbonyl group of the imide. In this case, the CT complex is an effective intermediate on the reaction pathway. In contrast, Brode[45] proposed that imidazole acts as a catalyst for the hydrolysis of thalidoimide (phthalimide-3-piperidinedione-2,6). The different results

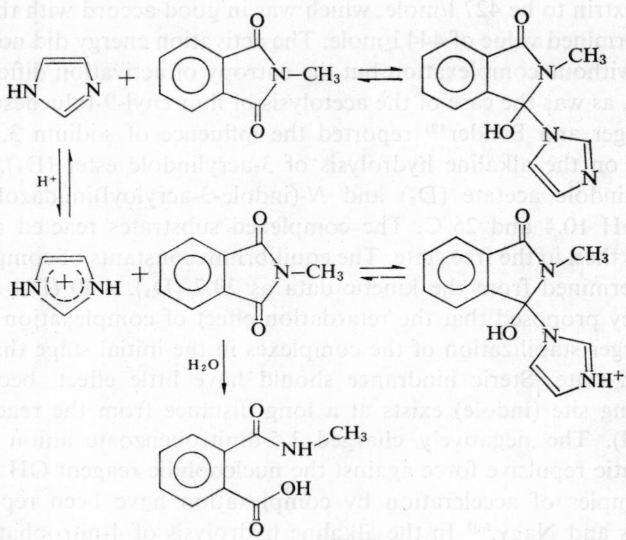

Fig. 3.20 Proposed mechanism of hydrolysis.

may be due to the formation of a tetrahedral intermediate (σ complex), as proposed by Su *et al.*

Kunitake and his co-workers[46] have investigated the hydrolysis of *p*-acetoxybenzoic acid in the presence of 4′-imidazole-4-naphthalene and 1-aminomethylnaphthalene, which exhibited catalytic behavior in this hydrolysis. They reported that the rate of the hydrolysis increased with increasing Michaelis-Menten constant K_m ($K_m = 1/K$). These naphthalene derivatives play an enzymatic role in the hydrolysis of carboxylic acids, probably through CT complexation. Letsinger and Klaus[47] have studied the catalytic action of poly-*N*-vinylimidazole on the hydrolysis of copolymers of acryl-*p*-vinylbenzoic acid-2,4-dinitrostyrene. They have postulated that the catalytic effect of imidazole derivatives is related to CT complex formation between the imidazole and the copolymer. The polyimidazole can catalyze the hydrolysis of *p*-nitrophenyl-polyuridylsuccinate about 300 times more rapidly than imidazole.

Lach and Chin[48] have reported comparatively large inhibition by β-cyclodextrin (heptapoly-α-D-glucopyranose) of the hydrolysis of benzocaine at 30°C ($Ba(OH)_2 = 0.04$ N). They supposed that the inclusion complex of benzocaine with β-cyclodextrin does not react; 1% β-cyclodextrin reduced the rate of hydrolysis to one-tenth, whereas a similar concentration of caffeine reduced it only 1.5 times. The constant of complex formation has been determined from the solubility of benzocaine in the presence of β-cyclodextrin to be 427 l/mole, which was in good accord with the kinetically determined value of 444 l/mole. The activation energy did not change with or without complexation but the entropy of activation differed considerably, as was the case of the acetolysis of fluorenyl-9-toluenesulfonate.

Menger and Bender[49] reported the influence of sodium 3,5-dinitro benzoate on the alkaline hydrolysis of 3-acrylindole ester (D_1), *p*-nitrophenyl-3-indole acetate (D_2) and *N*-(indole-3-acryloyl)imidazole amide (D_3) at pH 10.4 and 25°C. The complexed substrates reacted about 33 times less than in the free state. The equilibrium constants of complexation were determined from the kinetic data as 29.7 (D_1), 4.80 (D_2) and 21.5 (D_3). They proposed that the retardation effect of complexation was due to the larger stabilization of the complexes in the initial stage than in the transition state. Steric hindrance should have little effect, because the complexing site (indole) exists at a long distance from the reaction site (carbonyl). The negatively charged 3,5-dinitrobenzoate anion provides electrostatic repulsive force against the nucleophilic reagent OH^-.

Examples of acceleration by complexation have been reported by Bruylants and Nagy.[50] In the alkaline hydrolysis of 4-nitrophathalimide in an alcoholic solvent they observed twofold acceleration of the rate of hydrolysis in the presence of acenaphthene (5×10^{-5} mole/l). The stability constant of the 1:1 complex was 1.0 l/mole. Another example is that the aminolysis of 3,5-dinitrophthalic anhydride was enhanced in the presence of aromatic donor molecules such as anthracene and phenanthrene.

In summary, Table 3.10 presents the various effects of complexing donors and/or acceptors as third bodies in the hydrolysis of a series of cinnamates and acrylates.

In connection with the mechanism of coenzyme action, 1,3-dimethylalloxazine has been employed as a model compound of flavin mononucleotide (FMN), and its reactivity or its catalytic activity and the effects of EDA complex formation thereon have been examined by Iwasawa and Tamaru[51] for the redox process between methylene blue and ascorbic acid.

It was found that 1,3-dimethylalloxazine accelerates the redox reaction and that the formation of EDA complexes markedly affects the catalytic activity.

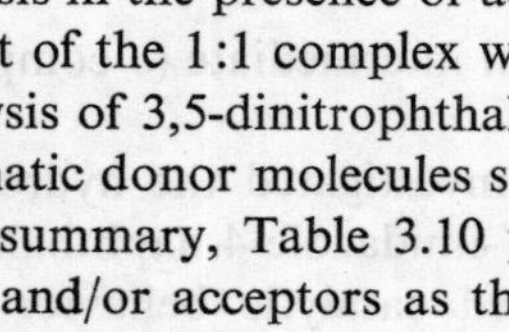

1,3-dimethylalloxazine 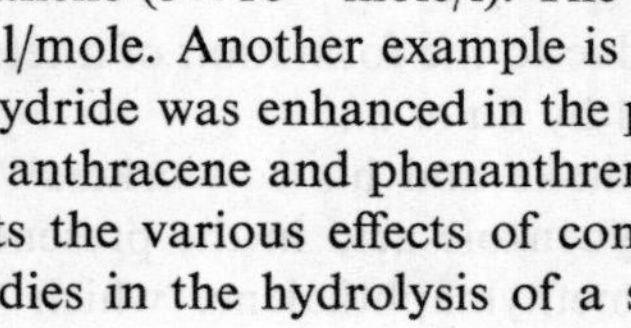 FMN

	8-Chlorotheophylline	25	[NaOH] = 0.1 N	0.08	22	—	—	[illegible]	(l)
	7-Theophylline acetic acid	25	—	0.06	14	—	—	1.6	(l)
	Theophylline (anion)	25	—	0.08	11	—	—	1.8	(l)
	Caffeine	25	—	0.03	18	—	—	[illegible]	(l)
	Guanine (anion)	25	—	0.06	<1	—	—	1.07	(l)
Cinnamoylsalicylic acid	Theophyllinate	25	pH = 12.80	—	8	—	—	16.6	(l)
	8-Chloro-theophylline	25	pH = 12.80	—	14	—	—	3.3	(l)
	8-Bromo-theophylline	25	pH = 12.80	—	15	—	—	2.9	(l)
Riboflavin	Caffeine	8	[NaOH] = 0.25 N	—	59.7	5.43 × 10[illegible]	1.0 × 10[illegible]	5.43	(m)
p-Nitrophenyl-benzoate	Theophylline	45	pH = 11.90	—	12	—	—	1.30	(l)
trans-Cinnamoyl-imidazole	Caffeine	25	pH = 1 ~ 2	0.09	—	—	—	1.22	(i)
Sodium-p-acetoxy-cinnamate	Theophylline	25	pH = 11.8 ~ 12	—	7.5	156.0	93.4	1.67	(n)
p-Acetoxybenzal-acetone	Theophylline	25	pH = 11.8 ~ 12	—	7	222.6	133.3	1.67	(n)
Methyl-p-acetoxy-cinnamate	Theophylline	25	pH = 11.8 ~ 12	—	8	224.0	112.0	2	(n)

(a) T. Higuchi and L. Lachman, J. Am. Pharm. Sci., 44, 521 (1955). (b) L. Lachman and T. Higuchi, ibid., 46, 32 (1957). (c) L. Lachman, L. J. Ravin and T. Higuchi, ibid., 45, 290 (1956). (d) L. Lachman, D. E. Guttman and T. Higuchi, ibid., 46, 36 (1957). (e) D. A. Wadke and D. E. Guttman, ibid., 53, 1073 (1964); 54, 1293 (1965). (f) J. L. Lach and T. F. Chin, ibid., 53, 924 (1964). (g) K. A. Connors and J. A. Mollica Jr., J. Am. Chem. Soc., 87, 123 (1965). (h) W. P. Jencks and M. Gilchrist, ibid., 90, 2622 (1968). (i) P. A. Kramer and K. A. Connors, ibid., 91, 2600 (1969). (j) S. C. K. Su and J. A. Shafer, J. Org. Chem., 34, 926 (1969). (k) A. Bruylands and J. B. Nagy, Bull. Soc. Chim. Belg., 75, 246 (1966). (l) J. A. Mollica Jr. and K. A. Connors, J. Am. Chem. Soc., 89, 308 (1967). (m) D. E. Guttman, J. Pharm. Sci., 51, 1162 (1962). (n) E. Brode, Arzneim. Forsch., 18, 1313 (1968).

(Source: O. B. Nagy and J. B. Nagy, Ind. Chim. Belg., 36, 940 (1971). Reproduced by kind permission of the Federation des Industries Chimiques de Belgique.)

The reduction of methylene blue by ascorbic acid was accelerated by a factor of 50 by the addition of DMA. The reaction was first-order with respect to methylene blue. The higher the temperature of the reaction, the greater the rate of the reaction, when the pH was constant.

1,2-Dichloroethane solutions of ascorbic acid, methylene blue, and a mixture of DMA–pyrene (2:1) were mixed at 0°C in the absence of oxygen. For 10 to 15 min the time-concentration relation for methylene blue exhibited a valley-shaped curve; thereafter, the reaction reached equilibrium. This valley curve can be explained not as an equilibrating reaction with a single reaction path, but as a successive reaction to consume methylene blue, followed by its regeneration.

Such characteristic behavior appeared only in the presence of both DMA and pyrene, and was not observed upon the addition of either of them or in phosphate buffer, where the interaction between them is very weak. Consequently, the following experiment was undertaken; first, ascorbic acid and methylene blue were reacted in the presence of only pyrene, and then DMA was introduced into the system during the course of the redox reaction. The results are given in Fig. 3.21. When the temperature of the reaction was kept constant at 26°C, the valley became shallower.

At low concentrations (molar ratios of DMA and pyrene to methylene blue of 1:22.5 and 1:45, respectively, or 1:45 and 1:90), no valley could be detected. When leucomethylene blue which had been obtained by bubbling hydrogen gas over a platinum catalyst in dichloroethane solution of methylene blue was introduced into a solution containing both DMA and pyrene, it was oxidized to methylene blue rapidly. DMA and pyrene form

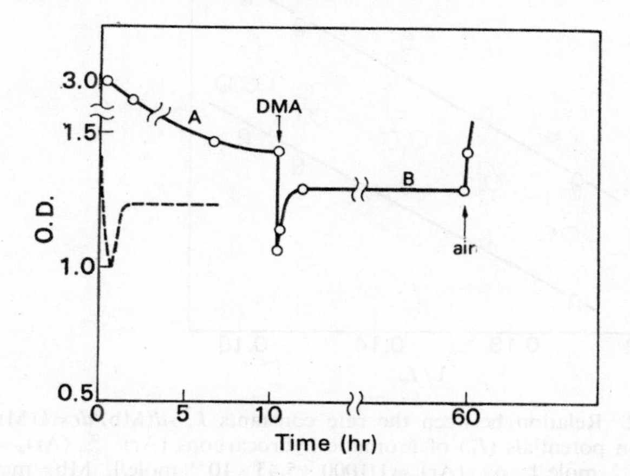

Fig. 3.21 Effect of EDA complex formation in the redox reaction of methylene blue with ascorbic acid at 0°C.[51) ---, In the presence of DMA and pyrene. A, Catalyzed by pyrene; B, DMA addition into A.

a solid EDA complex with a molar ratio of 2:1 and with a charge-transfer band in the region from 400 to 500 nm.[52] It was concluded, in accordance with these results, that the appearance of the characteristic curve is due to complex formation between DMA and pyene.

It was also confirmed spectroscopically and theoretically that enzymes or coenzymes form EDA complexes. When DMA, a model compound of FMN, formed an EDA complex with pyrene, there was a controlled reaction; this suggests that EDA complex formation plays an important role in the enzymatic reaction *in vivo*.

Various aromatic hydrocarbons, such as naphthacene, perylene, anthracene, pyrene, *p*-terphenyl, and phenanthrene, all electron donors for methylene blue but with different ionization potentials, accelerated the reduction of methylene blue by ascorbic acid, though they are generally considered to be inert in the reaction.

The stabilization energies for the formation of EDA complexes of aromatic hydrocarbons with methylene blue have been obtained by second-order perturbation theory, but if EDA complex formation makes a contribution to a decrease in the standard free energy of the activated state of the redox reaction, some correlation must exist between the stabilization energies and the rate constants. In fact, plots of the logarithms of the rate constant against the reciprocal of the ionization potential of aromatic hydrocarbons, a measure of the stabilization energy, gave a good linear relation, as shown in Fig. 3.22.

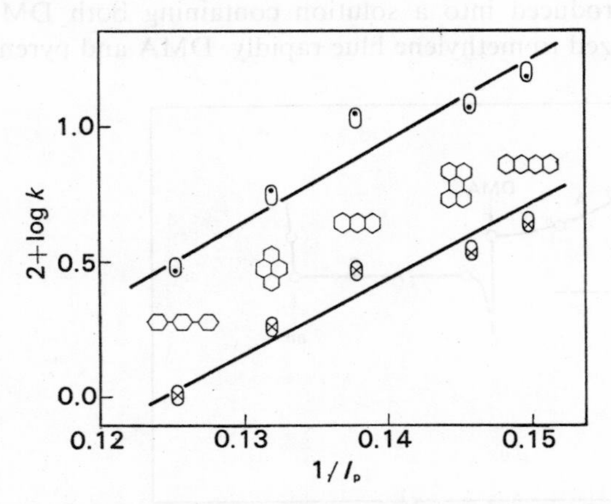

Fig. 3.22 Relation between the rate constants k, $-d(\mathrm{Mb})/dt=k(\mathrm{Mb})$, and ionization potentials (I_p) of aromatic hydrocarbons (Ar). ◎, $(\mathrm{Ar})_0=1/10\times 5.45\times10^{-5}$ mole/1; ⊗, $(\mathrm{Ar})_0=1/1000\times5.45\times10^{-5}$ mole/l. Mb=methylene blue.

REFERENCES

1. R. F. Jensen and H. C. Brown, *J. Am. Chem. Soc.*, **80**, 4046 (1958).
2. J. B. Ainscough and E. F. Caldin, *J. Chem. Soc.*, **1956**, 2528, 2540, 2546.
3. J. Meisenheimer, *Ann.*, **323**, 305 (1902); for infrared spectroscopic evidence that the Meisenheimer complexes isolated in solid form are covalently bonded and are not donor-acceptor interaction (CT complex) products, *see* L. K. Dyall, *J. Chem. Soc.*, **1960**, 5160; R. Foster and R. K. Meckie, *ibid.*, **1963**, 3796.
4. R. Destro, C. M. Gramaccioli and M. Scinonneta, *Nature*, **215**, 389 (1967); *Acta Cryst.*, **24**, 1369)1968).
5. M. J. Straus, *Chem. Rev.*, **70**, 667 (1970); E. Buncel, A. R. Norris and K. E. Russel, *Quart. Rev.*, **22**, 123 (1968).
6. C. R. Allen, A. J. Brook and E. F. Caldin, *Trans. Faraday Soc.*, **56**, 788 (1960); *J. Chem. Soc.*, **1961**, 2171.
7. T. Nogami, K. Yoshihara, H. Hosoya and S. Nagakura, *J. Phys. Chem.*, **73**, 2670 (1969); T. Nogami, K. Yoshihara and S. Nagakura, *Bull. Chem. Soc. Japan*, **44**, 295 (1971).
8. R. Foster and T. J. Thomson, *Trans. Faraday Soc.*, **58**, 860 (1962).
9. D. Buckley, S. Dunstan and H. B. Henbest, *J. Chem. Soc.*, **1957**, 4880.
10. G. A. Olah, S. J. Kuhn and S. H. Flood, *J. Am. Chem. Soc.*, **83**, 4571 (1961).
11. H. C. Brown and J. D. Brady, *ibid.*, **74**, 3570 (1952); L. J. Andrews and R. M. Keefer, *ibid.*, **79**, 1412 (1957).
12. S. Nagakura and J. Tanaka, *J. Chem. Phys.*, **22**, 563 (1954).
13. Z. Rappoport, *J. Chem. Soc.*, **1963**, 4498.
14. Z. Rappoport, *ibid.*, **1964**, 1348.
15. J. K. Williams, D. W. Wiley and B. C. McKusick, *J. Am. Chem. Soc.*, **84**, 2210 (1962).
16. D. W. Wiley and H. E. Simmons, *Progr. Phys. Org. Chem.*, **3**, 123 (1965).
17. S. Proskow, H. E. Simmons and T. L. Cairns, *J. Am. Chem. Soc.*, **84**, 2341 (1963).
18. P. Brown and R. C. Cookson, *Tetradedron*, **21**, 1977, 1993 (1965).
19. J. G. Martin and R. K. Hill, *Chem. Rev.*, **61**, 537 (1961).
20. N. C. Yang and Y. Gaoni, *J. Am. Chem. Soc.*, **86**, 5023 (1964); N. Wiberg and G. W. Buchler, *Chem. Ber.*, **97**, 618 (1964).
21. D. J. Cram and M. Goldstein, *J. Am. Chem. Soc.*, **85**, 1063 (1963).
22. E. M. Kosower, *ibid.*, **78**, 3497 (1956).
23. R. F. Powning and C. C. Kratzing, *Arch. Biochem. Biophys.*, **66**, 249 (1957); B. Ke, *ibid.*, **60**, 505 (1956).
24. M. B. Matthews and E. E. Conn, *J. Am. Chem. Soc.*, **75**, 5428 (1953).
25. Y. Prisse and G. Stein, *J. Chem. Soc.*, **1958**, 2905.
26. E. Euler, E. Alder and H. Hellström, *Z. Physiol. Chem.*, **243**, 239 (1936); E. Alder, H. Hellström and H. Euler, *ibid.*, **242**, 225 (1936).
27. M. B. Yarmolinsky and S. P. Colowick, *Biochim. Biophys, Acta*, **20**, 177 (1956).
28. E. M. Kosower and S. W. Bauer, *J. Am. Chem. Soc.*, **82**, 2191 (1960).
29. E. M. Kosower and E. J. Poziomek, *ibid.*, **85**, 2035 (1963).
30. W. V. E. Doering and W. E. McEwen, *ibid.*, **73**, 2104 (1951).
31. P. T. Lansbury and J. O. Peterson, *ibid.*, **85**, 2236 (1963).
32. E. M. Kosower and T. S. Sorensen, *J. Org. Chem.*, **27**, 3764 (1962).
33. O. B. Nagy and J. B. Nagy, *Ind. Chim. Bedg.*, **36**, 829, 929 (1971).

34. J. E. Leffler and R. A. Hubbard, *J. Org. Chem.*, **19**, 1089 (1954); B. B. Smith and J. E. Leffler, *J. Am. Chem. Soc.*, **77**, 2509 (1955); B. M. Graybill and J. E. Leffler, *J. Phys. Chem.*, **63**, 1461 (1959).
35. A. K. Colter and L. M. Clemens, *J. Am. Chem. Soc.*, **87**, 847 (1965).
36. A. K. Colter, S. S. Wang, G. H. Megerle and P. S. Ossip, *ibid.*, **86**, 3106 (1964).
37. A. K. Colter, F. F. Guzik and S. H. Hui, *ibid.*, **88**, 5754 (1966).
38. A. K. Colter and S. H. Hui, *J. Org. Chem.*, **33**, 1935 (1968).
39. K. Okamoto, I. Nitta and H. Shingu, *Bull. Chem. Soc. Japan*, **41**, 1433 (1968).
40. K. A. Connors and J. A. Mollica Jr., *J. Am. Chem. Soc.*, **87**, 123 (1965).
41. J. A. Mollica Jr. and K. A. Connors, *ibid.*, **89**, 308 (1967).
42. P. A. Kramer and K. A. Connors, *ibid.*, **91**, 2600 (1969).
43. K. A. Connors, M. H. Infeld and B. J. Kline, *ibid.*, **91**, 3597 (1969).
44. S. C. K. Su and J. A. Shafer, *J. Org. Chem.*, **34**, 926 (1969).
45. E. Brode, *Arzneim. Forsch.*, **18**, 1313 (1968).
46. T. Kunitake, S. Shinkai and C. Aso, *Bull. Chem. Soc. Japan*, **43**, 1109 (1970).
47. R. L. Letsinger and I. S. Klaus, *J. Am. Chem. Soc.*, **86**, 3884 (1964); *ibid.*, **87**, 3380 (1965); R. L. Letsinger and T. J. Saveveide, *ibid.*, **84**, 3122 (1962).
48. J. L. Lach and T. F. Chin, *ibid.*, **53**, 924 (1964).
49. F. M. Menger and M. L. Bender, *J. Am. Chem. Soc.*, **88**, 131 (1966).
50. A. Bruylants and J. B. Nagy, *Bull. Soc. Chim. Belg.*, **75**, 246 (1966).
51. Y. Iwasawa, M. Soma, T. Onishi and K. Tamaru, *Bull. Chem. Soc. Japan*, **43**, 720 (1970).
52. Y. Matsunaga, *Nature*, **211**, 182 (1966).

CHAPTER 4

Heterogeneous Catalysis by EDA Complex

4.1 Adsorption through EDA Complexing

The adsorption of gas molecules on solid surfaces has been considered in terms of charge-transfer phenomena. Mulliken[1] suggested a similarity between adsorption processes and charge-transfer interactions between gaseous donor or acceptor molecules and heterogeneous surfaces. Since that time, the charge-transfer theory has been applied by McGlymn.[2] to chemisorption on organic or inorganic semiconductors. One can understand heterogeneous catalytic reactions as chemical processes occurring through EDA complexation, i.e. "surface complex" formation between a solid surface acting as a *giant* donor or acceptor and the gaseous molecules.

The adsorption of hydrogen or oxygen gas causes marked changes of the surface potential of metals and of the numbers of charge carriers, probably due to charge-transfer interaction between the metal surface and adsorbed gas molecules. For instance, oxygen gas and carbon monoxide (electron acceptors) increase the electrical resistance, with the formation of partially charged adsorbed species. When electron-accepting gas molecules such as O_2, NO, NO_2 and N_2O approach *n*-type semiconductive metal oxides such as MgO and ZnO, delocalized electrons in conduction bands or localized electron pairs are transferred from the semiconductors to the adsorbed molecules, providing new chemical species such as O_2^-, N_2O^- and CO_2^-, which have been characterized by esr, ultraviolet and infrared spectrometry.[3] Hauffe *et al.*[4] have reported marked changes in dark or photoelectric conductivity due to the adsorption of donor or acceptor molecules onto organic semiconductors such as phthalocyanines and polycondensed aromatic hydrocarbons. They observed that the dark and photocurrents decreased on adsorption of NH_3, alkylamines, pyridine and

H_2O (electron donors) onto metal-free (H_2Pc) and Mg-phthalocyanines (MgPc), whereas they increased on adsorption of O_2 and NO (electron acceptors). Such conductivity changes occurred reversibly on the adsorption and desorption of these gas molecules. Besides oxygen and hydrogen, other gases also have characteristic reversible effects on the conductivity of β-carotine layers. As can be seen from Table 4.1, the conductivity changes of certain systems (e.g. β-carotene) range over six orders of magnitude, dependent on the type of gas.[5] As in the case of ZnO,[6] it seems feasible to utilize this effect for the detection of gases.

TABLE 4.1 Dependence of the dark conductivity of β-carotene on the nature of an adsorbed gas

Gas	$\sigma_{gas}/\sigma_{vac}$
NO	2
Butanol	7
Methyl acetate	10
Oxygen	10^2
SO_2	10^3
Acetone	10^4
Ethanol	10^5
NH_3	10^5
Methanol	10^6
NO_2	10^6

(After B. Rosenberg, T. Misra and R. Switzer, *Nature*, **217**, 423 (1968).)

Using a single crystal of anthracene, Waddington and Schneider[7] have observed a marked increase in photocurrent in the presence of BF_3, HCl, O_2 and NO (electron acceptors), and a decrease on the adsorption of donors such as NH_3, $(CH_3)_2O$ and $(CH_3)_3N$. The desorption proceeded slowly under higher vacuum conditions. In double-layer systems between acceptor and donor films (such as anthracene–I_2 and *p*-chloranil–NH_3), charge-transfer complex formation over the crystal surfaces caused marked effects on the bulk electrical conductivities. For instance, Kearns and Calvin[8] reported that when double-layer films were prepared by covering Mg-phthalocyanine film with oxidized *N*,*N*,*N'*,*N'*-tetramethyl-*p*-phenylenediamine, as shown in Fig. 4.1, the electrical conductivity of the Mg-phthalocyanine film increased and a considerable photovoltage was observed, with an accompanying esr signal at $g=2.000$ (Fig. 4.2). They also observed marked increases in the photoconductivities of metal-free phtha-

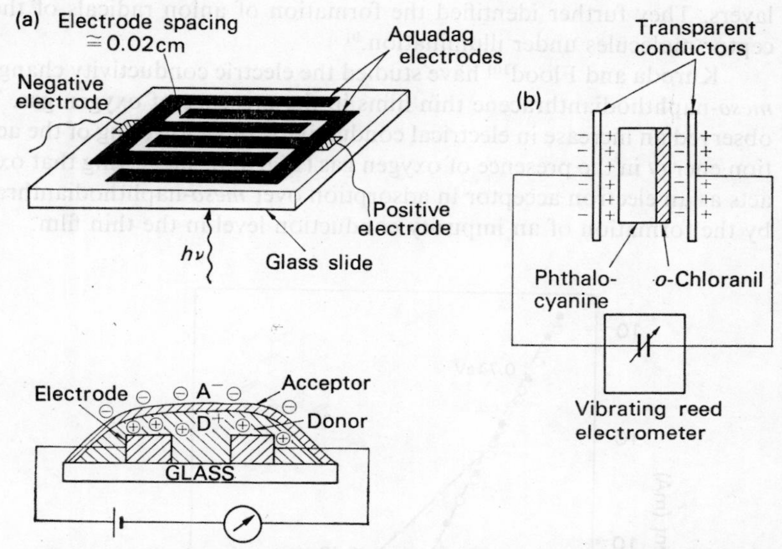

Fig. 4.1 (a) Layered donor-acceptor systems. (b) Schematic diagram of polarization apparatus.
(Source: M. Calvin, *Advan. Catalysis*, **14**, 11 (1963). Reproduced by kind permission of Academic Press, Inc., U.S.A.)

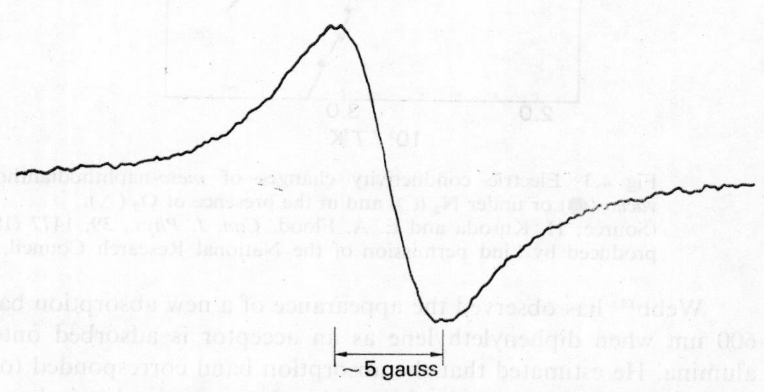

Fig. 4.2 Esr spectrum of *o*-chloranil-"doped" phthalocyanine. Curve represents the first derivative of absorption.
(Source: M. Calvin, *Advan. Catalysis*, **14**, 13 (1963). Reproduced by kind permission of Academic Press, Inc., U.S.A.)

locyanine and violanthene thin films in the presence of *o*-chloranil, I_2 or TCNE as strong acceptors, implying that complete charge transfer had occurred from the donor thin films to the acceptor molecules to form "*hole*" carriers in the donor thin films and electron carriers in the acceptor layers. They further identified the formation of anion radicals of the acceptor molecules under illumination.[9]

Kuroda and Flood[10] have studied the electric conductivity changes of *meso*-naphthodianthracene thin films in the presence of oxygen gas. They observed an increase in electrical conductivity and a lowering of the activation energy in the presence of oxygen gas (Fig. 4.3), suggesting that oxygen acts as an electron acceptor in adsorption over *meso*-naphthodianthracene by the formation of an impurity conduction level in the thin film.

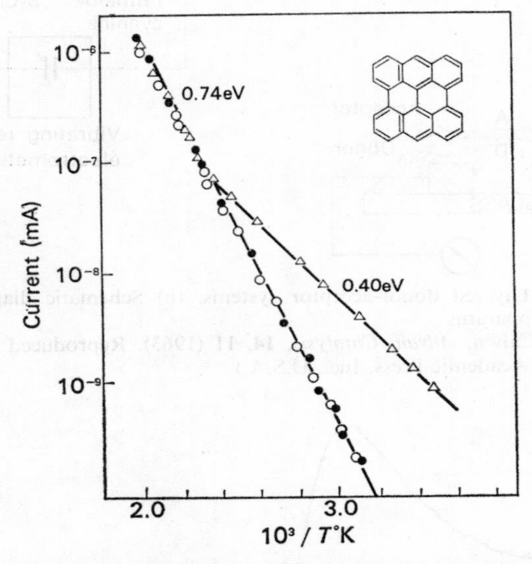

Fig. 4.3 Electric conductivity changes of *meso*-naphthodianthracene *in vacuo* (●) or under N_2 (○) and in the presence of O_2 (△).
(Source: H. Kuroda and E. A. Flood, *Can. J. Phys.*, **39**, 1477 (1961). Reproduced by kind permission of the National Research Council, Canada)

Webb[11] has observed the appearance of a new absorption band near 600 nm when diphenylethylene as an acceptor is adsorbed onto silica-alumina. He estimated that the absorption band corresponded to the CT band in the charge-transfer interaction between the Lewis base site of silica-alumina and diphenylethylene. Ross and Oliver[12] have also observed shifting of the absorption peaks of iodine (I_2) and pyridine (as a donor) during adsorption over alumina and silica-alumina surfaces. A new CT band appeared in the region of 350 nm. On adsorption of aromatic hydrocarbons such as perylene over the stronger Lewis acid sites of alumina and $AlCl_3$, Rooney and Pink[13] detected the formation of cation radicals of the corresponding aromatic donors by ultraviolet and esr spectrometry. Quali-

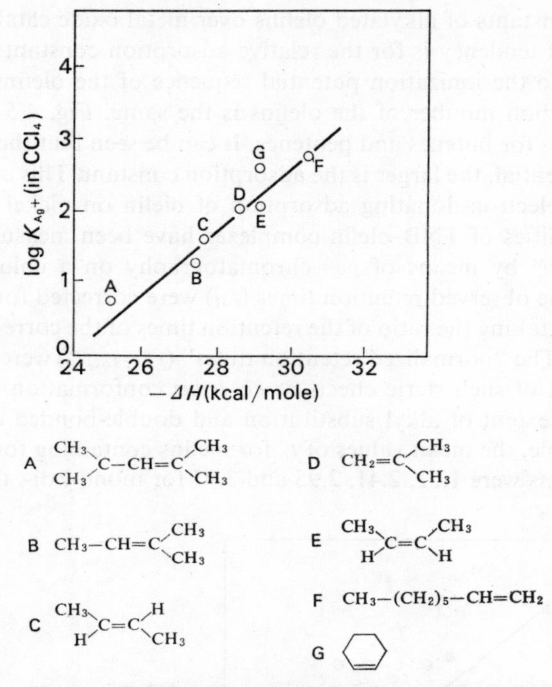

Fig. 4.4 Relation between the stability constants of Ag^+ complexes and the heats of hydrogenation of olefins.
(Source: R. D. Gardner, R. L. Brandon and N. G. Nix, *Chem. Ind.*, 1364 (1958). Reproduced by kind permission of the authors.)

tative and quantitative observations of spectral changes occurring on gas adsorption over a solid surface are useful tools for investigating the acidity and basicity of surfaces.

As shown in Fig. 4.4, Gardner, Brandon and Nix[14)] demonstrated a good correlation between the stability constants (K_X) of CT complexes (π complexes) of olefins with Ag^+ and the hydrogenation activities of the corresponding olefins over Pd and Ni metal surfaces. (Ag^+ is isoelectronic with Pd.)

$$R_2C=CH_2 + Ag^+ \overset{K_c}{\rightleftharpoons} \left[R_2C\overset{\downarrow}{=}CH_2 \atop Ag^+ \right]$$

A similar correlation has also been obtained by Morooka and Ozaki[15)] for the adsorption constants of alkylated olefins over metal oxide catalysts.

An important tendency is for the relative adsorption constants to be inversely related to the ionization potential sequence of the olefins, provided that the carbon number of the olefins is the same. Fig. 4.5 shows linear relationships for butenes and pentenes. It can be seen that the lower the ionization potential, the larger is the adsorption constant. This correlation suggests an electron-donating adsorption of olefin on nickel oxide. The relative stabilities of TNB–olefin complexes have been measured by Cvetanovič *et al.*[16] by means of gas chromatography on a column of TNB-firebrick. The observed retention times (t_{ol}) were corrected for physical interaction by taking the ratio of the retention times of the corresponding paraffins (t_p). The "normalized retention times" ($t_n = t_{ol}/t_p$) were found to be independent of such steric effects as *cis-trans* conformation and to increase with the extent of alkyl substitution and double-bonded carbon atoms. For example, the mean values of t_n for olefins containing four, five or six carbon atoms were 1.51, 2.41, 2.95 and 3.95 for mono-, di-, tri- and

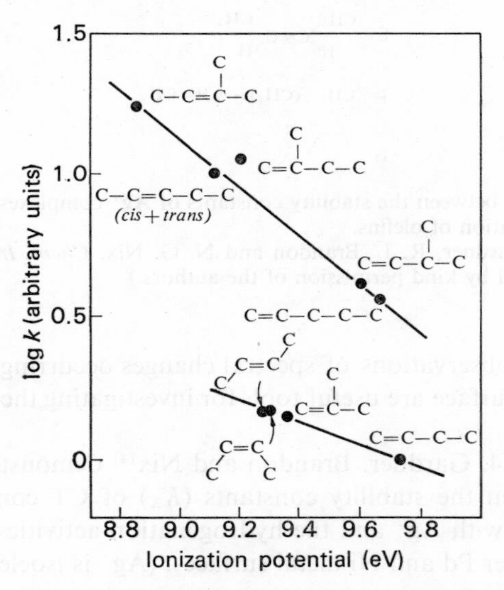

Fig. 4.5 Relation between the relative adsorption constants and ionization potentials of butene and pentene isomers. The ionization potential values were derived from (a) F. H. Field and J. L. Franklin, *Electron Impact Phenomena and the Properties of Gaseous Ions*, McGraw-Hill, 1957 (for 1-butene, 2-methylpropene and 2-methyl-2-butene), (b) R. E. Honig, *J. Chem. Phys.*, **16**, 105 (1948) (for *cis*- and *trans*-2-butene and 1-pentene), and (c) J. Collin and E. P. Lossing, *J. Am. Chem. Soc.*, **81**, 2064 (1959) (for 3-methyl-1-butene, 2-pentene and 2-methyl-1-butene).
(Source: Y. Morooka and A. Ozaki, *J. Am. Chem. Soc.*, **89**, 5127 (1967). Reproduced by kind permission of the American Chemical Society, U.S.A.)

tetrasubstituted olefins, respectively. These trends are in agreement with the trends of Morooka and Ozaki's results, so supporting the idea of electron-donating adsorption of olefins on nickel oxide. Thus, the adsorption complex during oxidation on nickel oxide catalyst may be a complex with some charge transfer. This model for adsorbed olefin is in accord with the observations of Enikeev *et al.*[17] during their measurements of work function changes upon adsorption of propylene. They observed a marked decrease in the work functions of semiconductor oxides such as nickel oxide, copper oxide, and vanadium pentoxide upon adsorption of propylene, and concluded that propylene acquired a partial positive charge when it was adsorbed on these oxides.

Nakamura and Otsuka[18] have observed a retarding effect of EDA interaction on the insertion reactions of organometallic hydride complexes. In the case of organic amines, charge-transfer interaction with an acceptor such as TCNE leads to products via "$\pi \rightarrow \sigma$" conversion of the complexes ($D^{\delta+} \ldots A^{-\delta} \rightleftharpoons D^+ - A^-$). The β-cyanoethylation of primary amines may also involve such charge-transfer interactions ($\pi \rightarrow \sigma$ conversion), which are not normally detectable because of rapid σ-bond formation (*cf.* Sect. 3.1). Nakamura and Otsuka, in contrast, have also found that the insertion of acrylonitrile into Cp_2MoH_2 (dicyclopentadienylmolybdenum dihydride complex) occurs at an unexpectedly slow rate, leading exclusively to the α-methylated product $[CP_2MoH(\underset{|}{C}H(CH_3)-CN)]$ via π-complex formation. Cp_2MoH_2 and Cp_2WH_2 readily form complexes with various strong electron-accepting olefins. Complexes of Cp_2WH_2 with π-acceptor olefins have characteristic CT bands (Table 4.2).

In the case of TCNE, the insertion reaction proceeds slowly at the M–H bond through rapid π-complex formation.

$$Cp_2\,MH_2 + \,\rangle C{=}C\langle\, \longrightarrow Cp_2\,M \begin{matrix} \diagup \overset{|}{C}-\overset{|}{C}-H \\ \diagdown H \end{matrix}$$

Since the dihydride metal complexes may be considered as a model system for the heterogenous active sites of catalysts, the insertion reaction of acceptor olefins into the M–H bond suggests the possible role of CT interactions in catalytic hydrogenation and insertion reactions over solid surfaces.

$$\begin{matrix} [D] & (M{-}H)^{+\delta}\ D & & \\ [A] & \underset{\displaystyle CN}{(\overset{}{C}H{=}CH_2)^{-\delta}}\ A & \xrightarrow{\text{slow } \pi\rightarrow\sigma \text{ conversion}} & \begin{matrix} M \\ | \\ CH{-}CH_3 \\ | \\ CN \end{matrix} \end{matrix}$$

TABLE 4.2 Maxima of CT bands in benzene

$[Cp_2MH_2] \cdot$ Acceptor

M=M$_o$ and W

Acceptor	Donors			
	Cp_2WH_2†		$C_6H_5N(CH_3)_2$	
	Color	ε_{max} (cm^{-1})	Color	ε_{max}(cm^{-1})
Maleic analydride	Blue-violet	17,500	Reddish-orange	24,600
Citaconic anhydride	Purple	19,200	Orange	26,200
Dimethylmaleic anhydride	Reddish-purple	~20,000	Orange	27,200
Fumaronitrile	Red	~20,000	Orange	27,000
m-Dinitrobenzene	Pale violet	17,800	Reddish-orange	23,800
1,3,5-Trinitrobenzene	Pale green	~16,800	Red	20,600

† 0.001~0.01 mmole/l with Cp_2WH_2 as a donor.

4.2 Activation of Molecular Hydrogen by EDA Complexes

Quantum theory led, in 1927, to the assertion that the hydrogen molecule can exist in two distinct and stable forms, *para*-hydrogen and *ortho*-hydrogen. In 1929, *para*-hydrogen was isolated and methods for the analysis of mixtures were developed. In 1931, the heavy hydrogen isotope deuterium was isolated, and catalytic chemists were thus able to study reactions considered to be of ideal simplicity. These reactions were:

$$p\text{-}H_2 \rightleftharpoons o\text{-}H_2 \tag{4.1}$$

and

$$H_2 + D_2 \rightleftharpoons 2HD \tag{4.2}$$

We would classify reaction (4.1) as a symmetry-forbiden reaction. Reaction (4.2), on the other hand, may be of a polar nature if the H–H bond is ruptured heterolytically (e.g. acid-base catalysts).

$$H_2 \rightleftharpoons H^+ + H^- \tag{4.3}$$

However, a free radical reaction may also occur if the H–H bond is ruptured homolytically (e.g. redox catalysts).

$$H_2 \rightleftharpoons \dot{H} + \dot{H} \tag{4.4}$$

A theoretical treatment of reaction (4.1) catalyzed by paramagnetic centers has been developed by Wigner.[19] The reaction was investigated using oxygen or nitric oxide in the gaseous phase, or oxygen, paramagnetic ions or free radicals in solution.

One interesting aspect of the reactions (4.1) and (4.2) is that on *ortho/para* conversion at a paramagnetic center, H–H bond may not be broken, and consequently H/D ($H_2 + D_2 \rightleftharpoons 2HD$) exchange may not occur. On the other hand, when H/D exchange does occur, a hydrogen-to-hydrogen bond is broken, and *ortho/para* conversion occurs as a result. Therefore, a comparison of the rates of the two processes should provide valuable information on the activation of molecular hydrogen by catalysts such as solid surfaces and complexes including EDA complexes. Activated charcoal, metal-phthalocyanines[20] and solid free radicals catalyze these reactions.

In Fig. 4.6, the rate of *ortho/para*-hydrogen conversion over dextrose charcoal is shown as a function of temperature; a definite minimum can be seen. This minimum was explained by assuming physical adsorption in the low temperature range (which decreases with rising temperature), and chemisorption in the high temperature range (which increases with rising temperature).

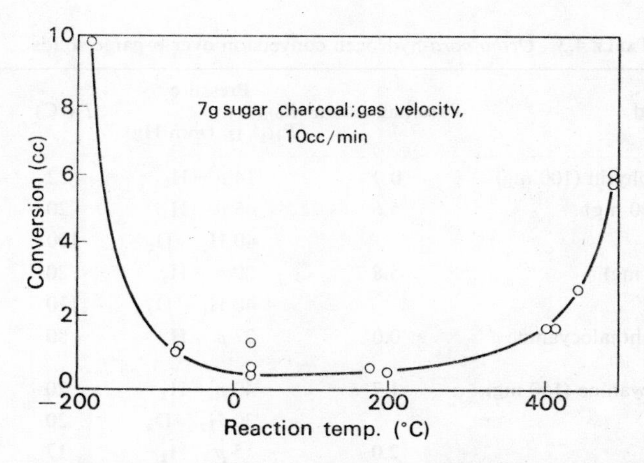

Fig. 4.6 Rate of *ortho/para* conversion of hydrogen over dextrose charcoal as a function of temperature.

Eley, Polanyi and Calvin[20,21] have comprehensively investigated the catalytic activities of porphines such as hematoporphyrin, hematin, hemin, metal-free phthalocyanine and copper-phthalocyanine for the *ortho/para*-hydrogen conversion and H_2–D_2 exchange reactions. Copper-phthalocyanine in the solid state was shown to catalyze *ortho/para* conversion at room temperature, but at temperatures up to 120°C no H_2–D_2 exchange was observed. Metal phthalocyanines other than Cu-phthalocyanine with no magnetic properties did not show *ortho/para* conversion or H/D exchange after 63 hr at −80°C. Paramagnetic porphine derivatives such as hematin and hemin did show catalytic activity for *ortho/para* conversion, but no H/D exchange occurred at up to 120°C, as shown in Table 4.3.

The stable free radical α,α'-diphenyl-β-picrylhydrazyl (DPPH) (Fig. 4.7) is a paramagnetic solid and has been studied extensively as a catalyst for *ortho/para* conversion. Harrison and McDowell[22] have reported that at liquid air temperature no hydrogen adsorption occurs and *ortho/para* conversion is very slow. The *ortho/para* conversion was found to proceed with first-order kinetics over solid DPPH in the temperature range between 90°K and 290°K, and showed a negative activation energy. More recent work by Eley and Inokuchi[23] confirmed that no H/D exchange occurs over DPPH as a solid or in solution, and adsorption of H_2 does not change the electrical conductivity of the solid, implying that no chemical bond is formed with molecular hydrogen. Development of the electron spin resonance (esr) technique has made it possible to correlate the *ortho/para* conversion activity with the measured paramagnetism of the solid.

The catalytic activities of Cu-phthalocyanine and two of its polymers

TABLE 4.3 *Ortho/para*-hydrogen conversion over organic solids

Organic solid	Paramagnetism	Pressure $P_{o/p-H_2}$ $P_{H_2-D_2}$ (mm Hg)	T (°C)	k (hr^{-1})
Hematoporphyrin (100 mg)	0.0	14 $p-H_2$	22	0
Hematin (200 mg)	5.6	65 $p-H_2$	20	0.036
		40 H_2-D_2	100	0
Hemin (100 mg)	5.8	20 $p-H_2$	20	0.011
		40 H_2-D_2	120	0
Metal-free phthalocyanine (150 mg)	0.0	27 $p-H_2$	−80	0
Cu-phthalocyanine (150 mg)	1.7	40 $p-H_2$	20	0.001
		30 H_2-D_2	20	0
DPPH	2.0	5 $p-H_2$	17	0.0026
		5 H_2-D_2	22	0

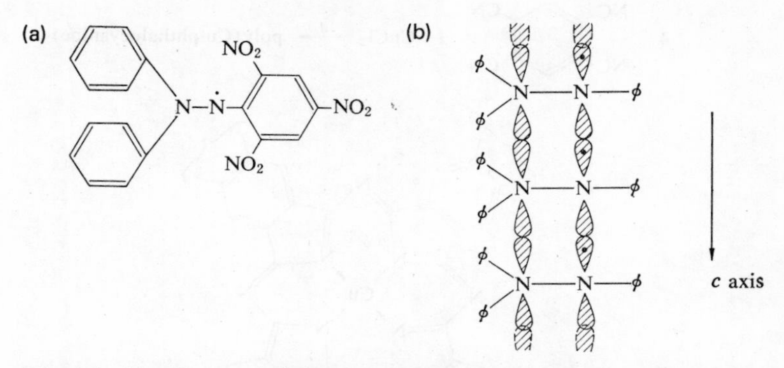

Fig. 4.7 (a) The α,α′-diphenyl-β-picrylhydrazyl (DPPH) molecule. (b) Crystal structure of solid DPPH.

for *ortho*/*para* conversion and H/D exchange have also been compared by Eley *et al.*[24] The activation energies for H_2–D_2 exchange were identical in the three materials, but differences in *ortho*/*para* conversion were found. Measurements were complicated, because the polymers (but not the monomer) tended to adsorb hydrogen in quantities equivalent to one atom of hydrogen for each copper atom present. The copper esr line disappeared after saturation with hydrogen. This phenomenon indicates that hydrogen diffuses easily throughout the entire polymer structure. The catalytic effects of organic materials cannot always be considered as surface phenomena, and care has to be taken when expressing the catalytic activity per unit surface area. The Cu-phthalocyanine polymer (Fig. 4.8) also exhibited a marked catalytic activity for H_2–D_2 exchange. The chemical activation process has an activation energy of about 10 kcal/mole (H_2–D_2 exchange), probably corresponding to activated π electrons in the polymerized conjugated carbon skeleton, since this activity is most marked in poly(Cu-phthalocyanine). The D_2–HZ exchange reaction with solid poly(Cu-phthalocyanine) takes place at 523°K, with an activation energy of about 23 kcal/mole. This activation process may perhaps be correlated with unpaired electrons, electrons formed by rupture of –CH and C-C bonds, etc. during heat treatment and polymerization, giving rise to the observed narrow esr line with a g value of around 2.000, due to diradical formation

$$-\text{CH}{=}\text{CH}- \xrightarrow{\Delta} -\!\!\left(\overset{\cdot}{\text{C}}\text{H}-\overset{\cdot}{\text{C}}\text{H}\right)\!\!-$$

in the conjugate system.

Earlier in this section it was mentioned that *ortho*/*para* conversion may occur at a paramagnetic center, while H/D exchange may occur via the homolytic or heterolytic splitting of the H-H bond. We may form a

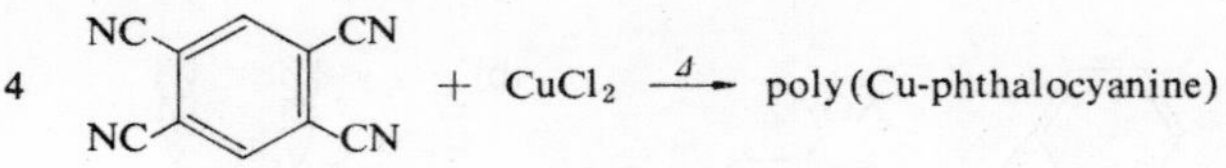

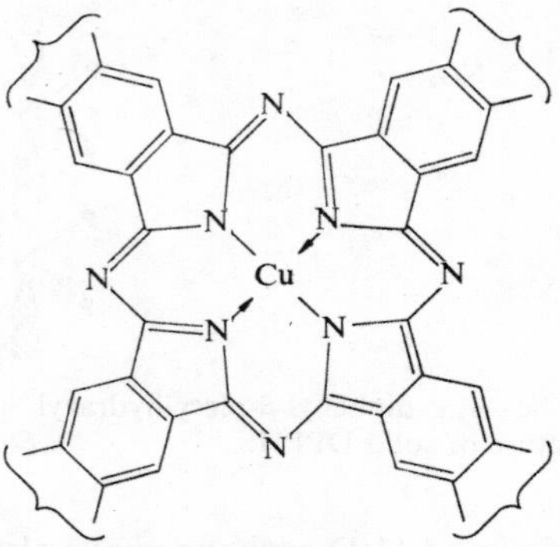

$$4\ C_6H_2(CN)_4 + CuCl_2 \xrightarrow{\Delta} \text{poly(Cu-phthalocyanine)}$$

Fig. 4.8 Poly(Cu-phthalocyanine).

paramagnetic center as well as a polar center by treating an electron donor with an electron acceptor, forming EDA complexes. This can be illustrated as follows by the simple case of the reaction between sodium metal and naphthalene.

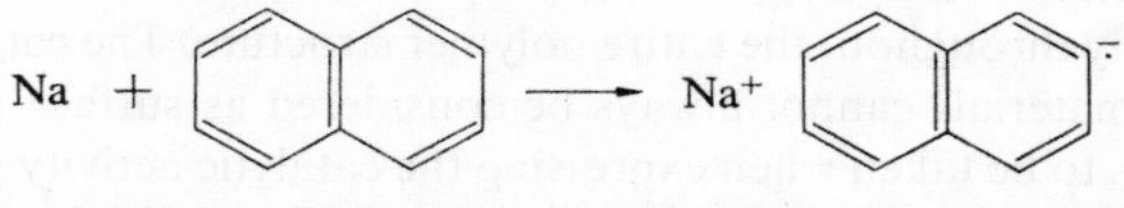

$$Na + C_{10}H_8 \longrightarrow Na^+ \ (C_{10}H_8)^{\dot{-}}$$

With a paramagnetic and a polar center together, it is not surprising that these complexes are active catalysts for the activation of molecular hydrogen. Systems of this type have been extensively studied by Japanese workers.

If Cs metal vapor is brought into contact with layers of tetracyanopyrene, a deep violet complex is formed. Kondow and Inokuchi[25] have studied the activity of films of this violet-colored EDA complex of *ortho/para* conversion and H/D exchange over a wide range of temperatures. They found that at 77°K *ortho/para* conversion occurs on this complex, but H/D exchange does not. Between 190° and 373°K, *ortho/para* conversion and H_2–D_2 exchange occur at comparable rates and with the same activation energy (3.7 kcal/mole), as shown in Fig. 4.9. Neither reaction is catalyzed by Cs metal or tetracyanopyrene alone up to 120°C. The *ortho/para* reaction occurring at 77°K is a physical one (having a negative activation energy, as shown in Fig. 4.9) and depends upon purturbation of the nuclear spin of the hydrogen molecules by the inhomogeneous magnetic field at the complex surface.

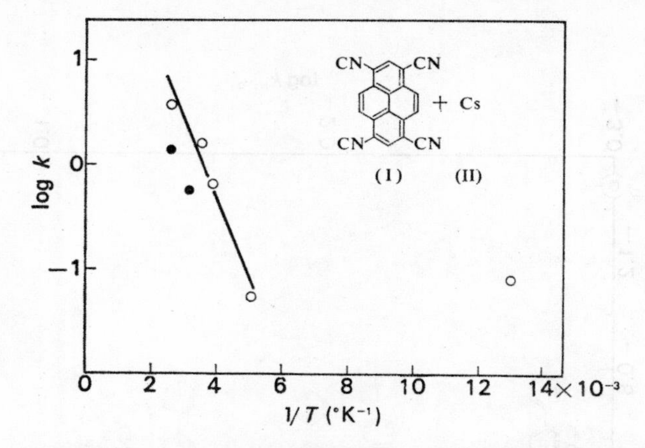

Fig. 4.9 The rate of conversion at various temperatures, for H/D exchange (●) and *ortho/para* hydrogen conversion (○).
(Source: T. Kondow, H. Inokuchi and N. Wakayama, *J. Chem. Phys.*, **43**, 3776 (1965). Reproduced by kind permission of the American Institute of Physics, U.S.A.)

In more extensive studies by Inokuchi, Mori and Wakayama[26] and by Ichikawa, Soma, Onishi and Tamaru,[27,28] it was found that two mechanisms of exchange could be distinguished over aromatic compound–alkali metal complex films:

$$H_2 + D_2 \rightleftharpoons 2HD \text{ (chemisorption)} \tag{4.5}$$

and

$$D_2 + HZ \rightleftharpoons HD + DZ \text{ (incorporation of D into the complex HZ)} \tag{4.6}$$

Occurrence of the mechanisms (4.5) and (4.6) depends upon the chemical structure of the electron acceptor and/or alkali metal, and also upon the ratio between the metal and electron acceptor.

The dependence of the rate of H/D exchange on the electronic properties of the acceptor molecules has been studied extensively by Ichikawa, Soma, Onishi and Tamaru.[28] Various types of aromatic acceptors have been used to prepare EDA-complex films with alkali metals such as sodium, and the catalytic activities of the complex films for H_2–D_2 exchange via chemisorption have been measured approximately under similar reaction conditions. The relative first-order rate constants and activation energies for the H_2–D_2 exchange reaction at 80°C are plotted in Fig. 4.10

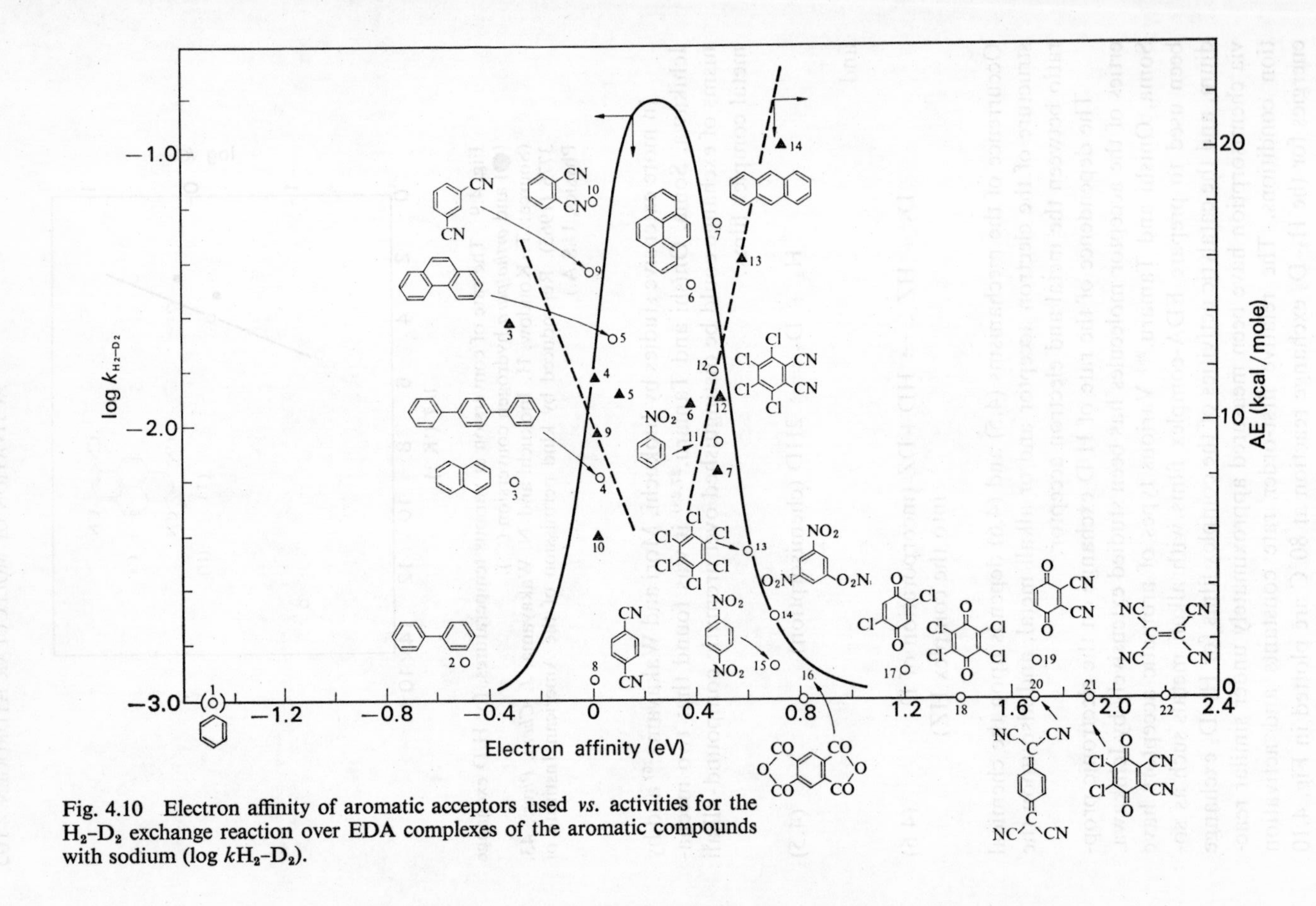

Fig. 4.10 Electron affinity of aromatic acceptors used *vs.* activities for the H_2–D_2 exchange reaction over EDA complexes of the aromatic compounds with sodium (log kH_2–D_2).

against the electron affinity (EA) (as estimated by Briegleb[29]) of the acceptor molecules in the EDA complexes. It is of interest to note that the activity of EDA complexes is strongly dependent upon the electron affinity of the acceptor molecules: the activity has a maximum in the region of 0 to 0.6 eV electron affinity of the acceptor molecules and is very small or negligible for stronger or weaker acceptor molecules.

4.2.1 Hydrogen adsorption

EDA complexes of aromatic compounds with alkali metals adsorb considerable amounts of hydrogen at room temperature in the solid state as well as in solution. The adsorbed hydrogen can be removed from the solid complexes *in vacuo*. After hydrogen adsorption, the paramagnetic spin concentration ($g=2.000$) of complex films such as anthracene–Na,[30] tetracyanopyrene–Cs and benzonitrile–Na decreases considerably.

Inokuchi, Wakayama and Hirooka[31] have investigated the effect of hydrogen adsorption on the electrical conductivity of perylene–Cs complex films. The electrical resistivity of the complex was semiconductive in character ($f=f_o \exp \varepsilon/kT$), as shown in Fig. 4.11. The activation energy was 0.08 eV. The admission of purified hydrogen caused a remarkable rise in resistance. The effect was reversible, since on pumping the hydrogen out the resistance of the cell was found to return almost to its original value. The estimated activation energy for conductivity of the film in the presence of hydrogen was 7 kcal/mole (0.3 eV). On the basis of these results, the abrupt change in electrical resistivity may thus correspond to the chemisorption of hydrogen onto the complex. Inokuchi *et al.*[31] suggested the formation of some kind of covalent link between the complex and dissociated hydrogen atoms, the latter immobilizing the conduction electrons. This assumption is supported by the detection of nmr absorption in a graphite–Cs complex on which hydrogen had been adsorbed.[32]

Tanaka *et al.*[30] have investigated the mechanism of hydrogen adsorption in the solid state as well as in solution by employing stoichiometric 1:1 and 1:2 anthracene–alkali metal EDA complexes. When hydrogen gas was admitted onto 1:1 and 1:2 anthracene–sodium complex films ($An^{-}Na^{+}$ and $An^{2-}2Na^{+}$), which were prepared from the corresponding complex solutions by evaporating the solvent, a considerable amount of hydrogen was adsorbed at 50–100°C, as illustrated in Fig. 4.12. Hydrogen easily diffused throughout the complex solids, and at saturation, one mole of the complex An^{2-} $2Na^{+}$ had adsorbed about one mole of hydrogen. Accompanying hydrogen adsorption, the characteristic absorption peaks of the mono-anion (at 720 nm) and dianion of anthracene (at 620 nm) decreased

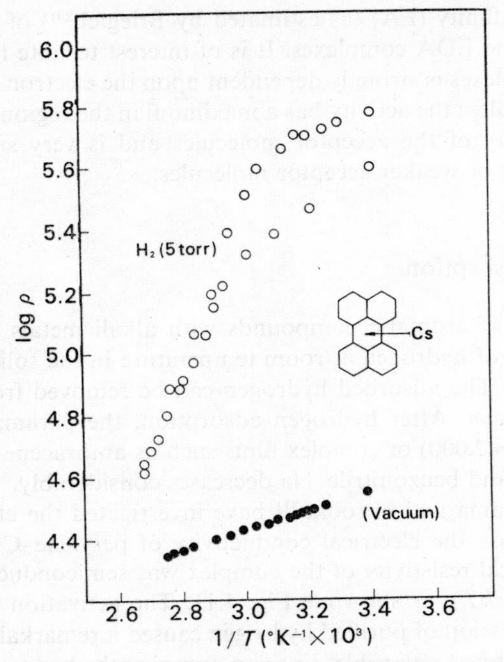

Fig. 4.11 Temperature dependence of the electric conductivity of the perylene–Cs complex *in vacuo* (●), and 5 mm Hg of H_2 (○). The specific resistivity of the complex *in vacuo* is *ca.* $10^0 \Omega$ cm at room temperature. (Source: H. Inokuchi, N. Wakayama and T. Hirooka, *J. Catalysis*, **8**, 383 (1967). Reproduced by kind permission of Academic Press, Inc., U.S.A.)

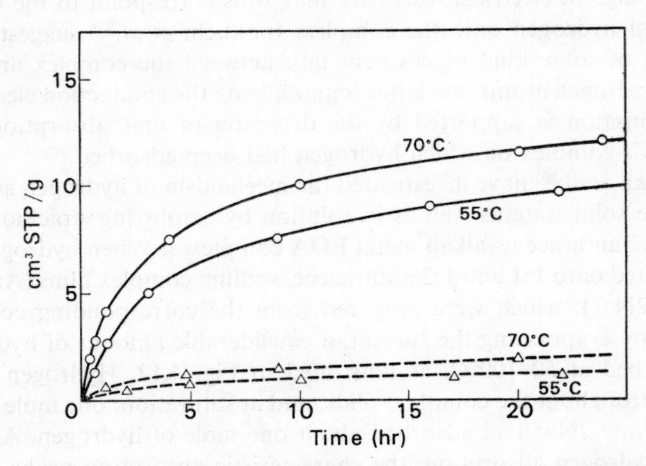

Fig. 4.12 Adsorption of hydrogen (initial perssure, 25 cm Hg) on films of $An^{2-}2Na^+$ (○) and An^-Na^+ (△) at 55 and 70°C.

in intensity and a new absorption peak appeared in the region of 420 nm in the solid state (432 nm in solution)*. Similar hydrogen absorption also occurred at room temperature in stoichiometric complex solutions, producing a new absorption peak at 432 nm. The absorption peak at 432 nm can be attributed to the formation of 9-monohydroanthracene anion salt, which can be independently prepared by the reaction between $An^{2-}2M^{+}$ and 9,10-dihydroanthracene (AnH_2).

$$An^{2-} \; 2M^{+} + AnH_2 \rightleftharpoons 2\,AnH^{-}M^{+}$$

M=Li, Na, K, Rb, Cs

On the basis of theoretical studies it is accepted that the spectra for carbanions and carbonium ions of alternant hydrocarbons are very similar. The spectra of many carbonium ions have been studied in detail by Dallinga *et al.*[33] The carbanion which has an absorption band at 398 nm in the ($An^{2-}2Li^{+}+AnH_2$) system, according to Velthorst *et al.*,[34] may be attributed to the 9-monohydroanthracene anion (AnH^{-})

Schneider has also shown by nmr measurements[35] that the CH_2 group in this carbanion has an aliphatic character. The nmr spectra of the 9-monohydroanion salt with different alkali metal cations have been observed and the coupling constants of non-equivalent protons estimated (Fig. 4.13).

In addition, when $An^{2-}2Na^{+}$ complex which had adsorbed hydrogen at 95°C was dissolved in THF or dimethoxyethane, a new signal ($\tau=8.73$), which is the same as that of sodium hydride, was observed. Consequently, it was concluded that hydrogen molecules were heterolytically dissociated into H^{+} and H^{-} over $An^{2-}2Na^{+}$ complex to form $AnH^{-}Na^{+}NaH$:

$$An^{2-} \; 2Na^{+} + H_2 \rightleftharpoons AnH^{-} \; Na^{+}\cdot NaH$$

* By evacuating the complex film at 80–120°C, the original absorption peak at 620 nm was completely restored.

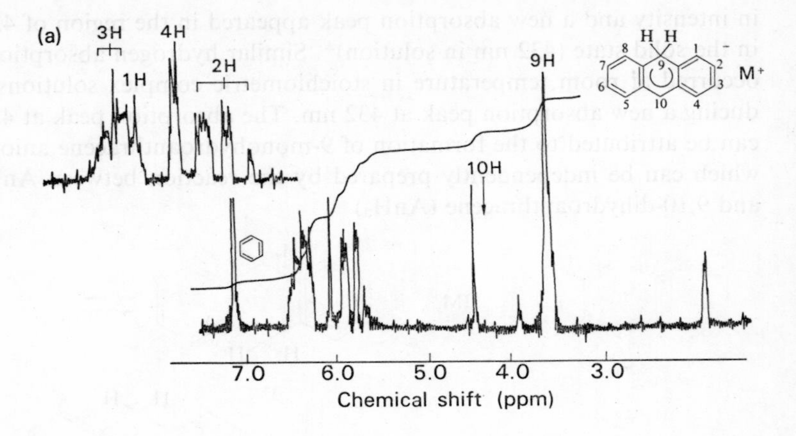

(b)

	Electronic spectrum maximum absorption peak (nm)	Chemical shift (ppm)			
		9-H	10-H	4-H	1-H
AnH-Li$^+$	375	3.65	4.35	6.08	6.42
AnH-Na$^+$	432	3.68	4.44	6.30	6.43
AnH-K$^+$	450	3.71	4.54	6.19	6.46
AnH-Rb$^+$	500	—	—	—	—
AnH-Cs$^+$	490	—	—	—	—

Fig. 4.13 Nmr spectra of 9-monohydroanthracene anion complex with different alkali metal cations. (a) 9-Monohydroanthracene–K salt deuterated with D_2 in [$^2H_8$]-THF (benzene-locking, 35°C, 100 MHz). (b) Data for monohydro-anion salts in [$^2H_8$]-THF (benzene-locking, 100 Hz).

The rate of adsorption and amount of adsorbed hydrogen increased with temperature. The activation energy for hydrogen adsorption was about 13 kcal/mole over the $An^{2-}2Na^+$ complex, and the rate and amount of hydrogen adsorption over the $An^{2-}2Na^+$ complex were each 10 times greater than those over the An^-Na^+ complex. In the case of the An^-Na^+ complex, hydrogen adsorption similarly yielded 9-monohydroanthracene anion salt and sodium hydride with one mole of neutral anthracene:

$$[\text{anthracene}]^- \, Na^+ + H_2 \rightleftharpoons$$

$$[\text{9-monohydroanthracene (H H)}]^- \, Na^+ \, NaH + [\text{anthracene}]$$

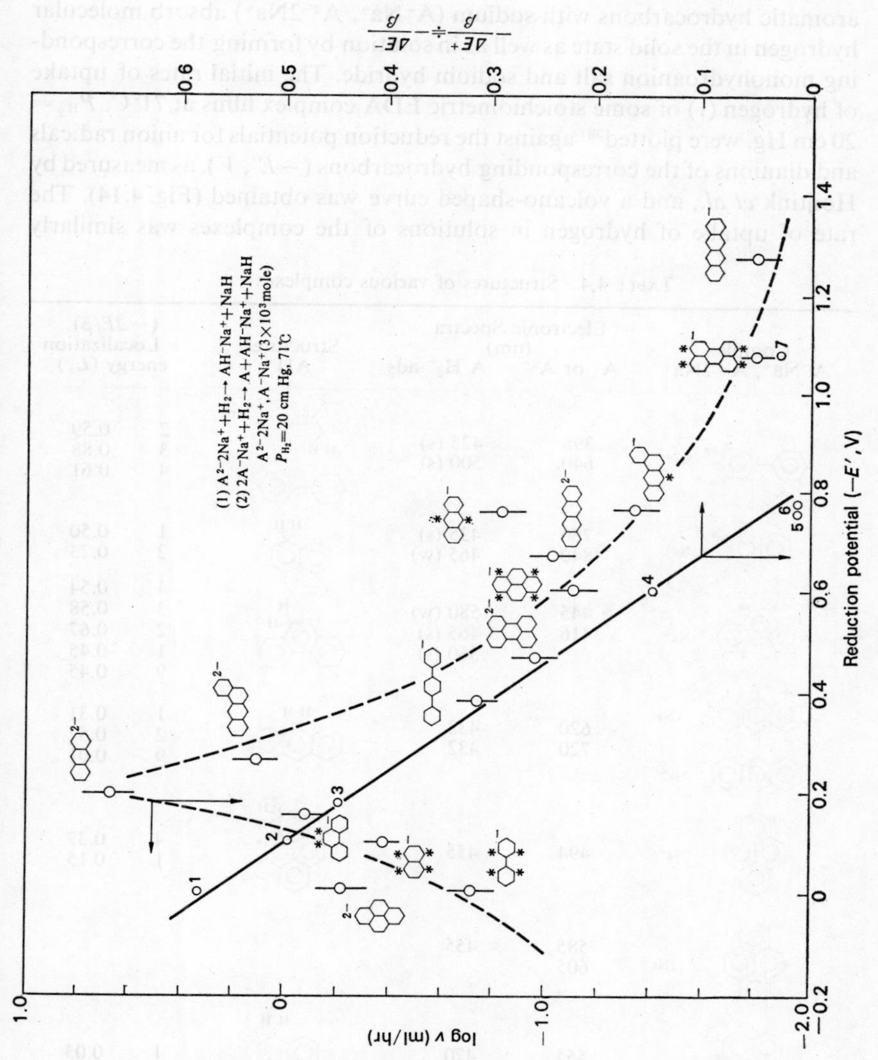

Fig. 4.14 Initial rate of H_2 adsortion (ν) *vs.* reduction potential.

The concentration of unpaired electrons in the An^-Na^+ complex film decreased considerably after hydrogen adsorption at 95°C.

In general, the stoichiometric 1:1 and 1:2 EDA complexes of various aromatic hydrocarbons with sodium (A^-Na^+, $A^{2-}2Na^+$) absorb molecular hydrogen in the solid state as well as in solution by forming the corresponding monohydroanion salt and sodium hydride. The initial rates of uptake of hydrogen (v) of some stoichiometric EDA complex films at 71°C, P_{H_2}= 20 cm Hg, were plotted[36] against the reduction potentials for anion radicals and dianions of the corresponding hydrocarbons ($-E'$, V), as measured by Hoijtink *et al.*, and a volcano-shaped curve was obtained (Fig. 4.14). The rate of uptake of hydrogen in solutions of the complexes was similarly

TABLE 4.4 Structures of various complexes

Complex A^-Na^+, $A^{2-}2Na^+$	Electronic Spectra (nm) A^- or A^{2-}	A H_2^- ads	Structure AH^-	($-\Delta E/\beta$) π Localization energy (L_π)	
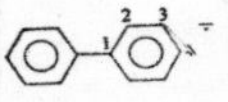 Na^+	398 640	425 (s) 500 (s)	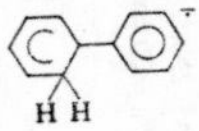 and 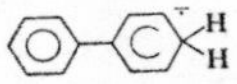	2 3 4	0.59 0.88 0.61
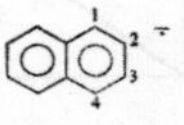 Na^+	760 840	435 (s) 465 (w)	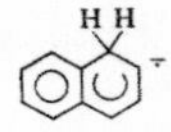	1 2	0.50 0.73
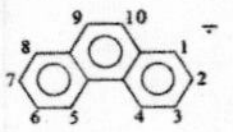 Na^+	445 416	580 (w) 465 (s) 360 (s)	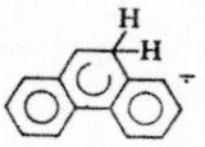	4 3 2 1 9	0.54 0.58 0.67 0.45 0.45
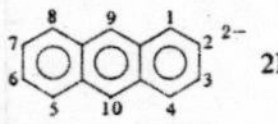 $2Na^+$ 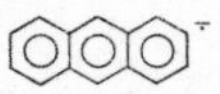Na^+	620 720	432 432	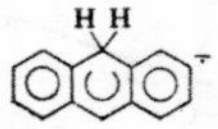	1 2 9	0.31 0.57 0.01
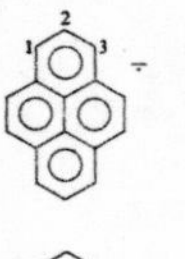 Na^+	494	455	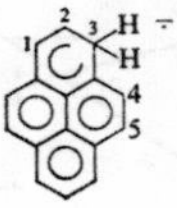	4 1	0.37 0.15
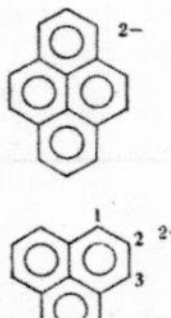 $2Na^+$	585 605	455			
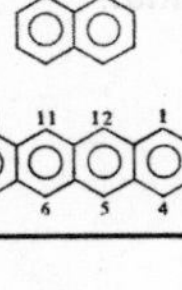 $2Na^+$	555 578	470	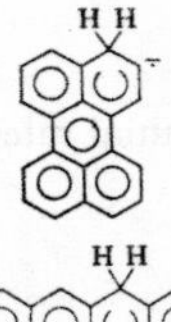	1 3	0.03 0.13
$2Na^+$					

correlated with their reduction potentials. Such a correlation between hydrogen adsorption by EDA complexes and reduction potential is probably analogous to that between the activities for H_2–D_2 exchange and the electron affinity of aromatic acceptors in EDA complexes.[28,37)]

When hydrogen gas was introduced into solutions of biphenyl–, naphthalene–, phenanthrene– and anthracene–sodium (1:1 and 1:2) at 20°C, it was found spectroscopically that new peaks appeared rapidly soon after the introduction of hydrogen (at (425, 500), (435, 465), (360, 465) and (432) nm, respectively), and were attributed to the formation of the corresponding monohydroanions, AH^-Na^+, as summarized in Table 4.4. Hydrogen molecules appear to be activated by heterolytic splitting, as in anthracene–Na complexes.

The stabilities of the monohydroanions (AH^-) may be related to the localization energy of the aromatic hydrocarbons. In Fig. 4.14, the localization energies ($-\Delta E^-/\beta$) calculated by the SCF-MO method are also given, in units of β. The positions where electrons are localized in each aromatic anion are indicated by asterisks. The results suggest that the hydrogen molecule reacts with aromatic anions having strong electron-donating ability and appropriate localization energy, resulting in the formation of a monohydroanion and/or hydride ion.

Ichikawa and Tamaru[38)] have measured hydrogen absorption and hydrogen exchange with various alkali metal donors in these complexes. D_2 gas (20 cm Hg) was admitted into solutions of the 1:2 complexes of anthracene with sodium and with potassium and the uptake of deuterium gas (absorption) and HD formation in the HZ–D_2 exchange reaction were measured at various temperatures. The results are given in Fig. 4.15, and imply that the uptake of hydrogen proceeded stoichiometrically in the 1:2 anthracene–Na complex with slow HD formation, whereas in the 1:2 anthracene–K complex, reversible incorporation of the hydrogen of the anthracene dianion (9-position) into D_2 occurred without detectable hydrogen uptake.

Ichikawa and Tamaru also observed[38)] the rates and activation energies for D_2 absorption and exchange in solutions of anthracene EDA complexes with various alkali metals, as shown in Table 4.5. The notable point is that hydrogen absorption activity is very high in solutions of lithium and sodium complexes, whereas a reversible exchange reaction proceeds without hydrogen absorption in the potassium, rubidium and cesium complex solutions. Although quite different behavior as regards hydrogen activation is observed with different alkali metals, the total activities (i.e. the sums of the rates of hydrogen absorption and reversible HZ–D_2 exchange) decrease monotonically with increasing ionic radius of the alkali cations used.

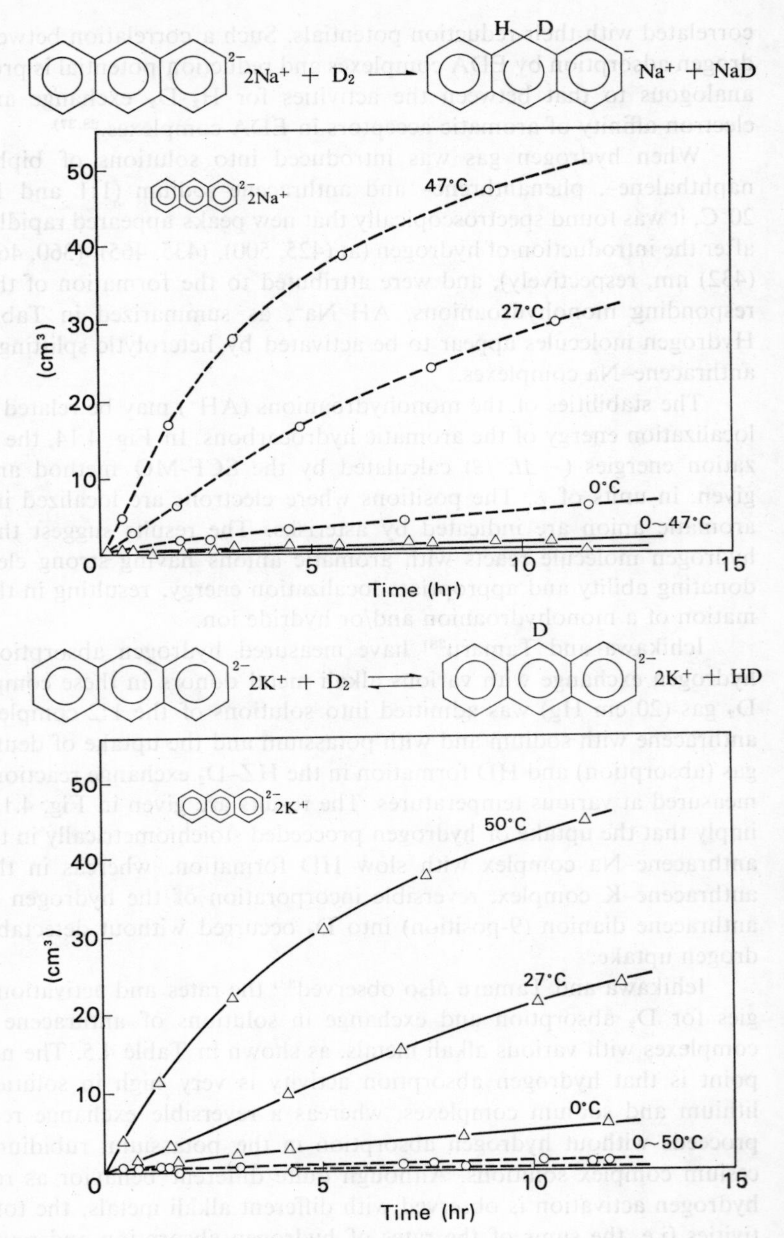

Fig. 4.15 Hydrogen absorption and HZ–D_2 exchange in solutions of anthracene^{2-}2Na^+ and anthracene^{2-}2K^+ (HZ). Complex, 6×10^{-3} mole; P_{D_2}, 24 cm Hg; solvent, DME. ○, D_2 absorption; △, HD formation.

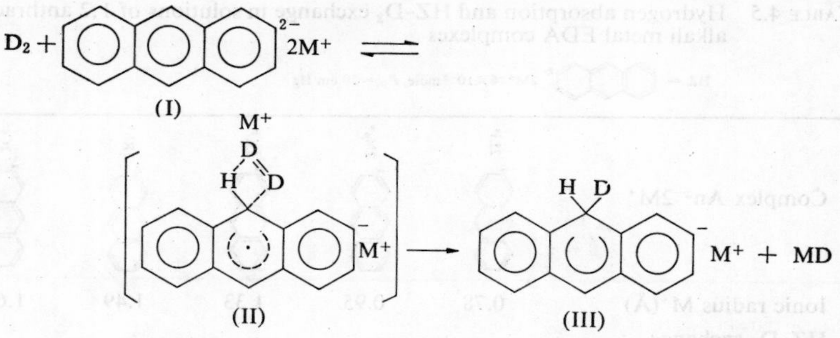

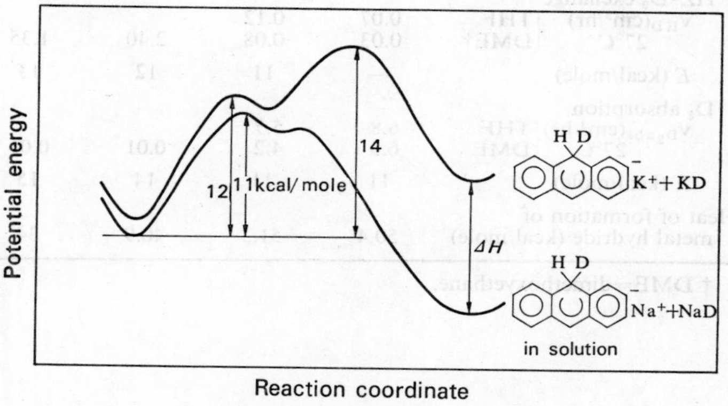

Heats of formation of metal hydrides

ΔH	LiH	NaH	KH	RbH	CsH
(kcal/mole)	50.4	51.3	40.9	34	30

Fig. 4.16 Energy profiles for hydrogen activation by 1:2 EDA complexes of anthracene with sodium and potassium.

It is possible to draw energy profiles for hydrogen activation by the 1:2 anthracene EDA complexes with sodium and potassium (Fig. 4.16). In the lithium and sodium complex solutions, the hydrogen molecule is dissociated through a transition state (II) to form a 9-monohydroanthracene anion salt and metal hydride as stable products, probably due to the large stabilization energy of the corresponding metal hydrides. The heat of formation of each alkali metal hydride is listed in the lower part of Fig. 4.16. Reversible hydrogen exchange proceeds only through the reversible steps (I) $\rightleftharpoons$ (II) in the K, Rb and Cs complex solutions, not proceeding to form the monohydroanion salt and hydride, due to the smaller stabilization of hydride ion with these alkali cations (Table 4.5).

TABLE 4.5 Hydrogen absorption and HZ–D_2 exchange in solutions of 1:2 anthracene–alkali metal EDA complexes

HZ = (anthracene)$^{2-}$ $2M^+$: 6×10^{-3} mole, P_{D_2} = 20 cm Hg

Complex $An^{2-}2M^+$	$2-$ $2Li^+$	$2-$ $2Na^+$	$2-$ $2K^+$	$2-$ $2Rb^+$	$2-$ $2Cs^+$
Ionic radius M^+(Å)	0.78	0.95	1.33	1.49	1.69
HZ–D_2 exchange					
v_{HD}(cm^3/hr) 27°C THF	0.07	0.12			
DME†	0.03	0.08	2.40	1.35	0.58
E (kcal/mole)	—	11	12	13	13
D_2 absorption					
v_{D_2abs}(cm^3/hr) 27°C THF	6.8	5.7			
DME	6.3	4.2	0.01	0.01	0.03
E (kcal/mole)	11	11	14	15	—
Heat of formation of metal hydride (kcal/mole)	50.4	51.3	40.9	34	30

† DME=dimethoxyethane.

4.2.2 The H_2–D_2 and HZ–D_2 hydrogen exchange reactions

Ichikawa and Tamaru[39,40] have presented a correlation between the hydrogen-exchange activity and structural properties of aromatic anions of EDA complexes. They prepared solutions of 1:1 and 1:2 EDA complexes of various aromatic hydrocarbons with potassium in dimethoxyethane. The reversible HZ–D_2 exchange reaction took place without hydrogen absorption. Deuteration occurred at the carbon atoms having the smallest localization energy. The asterisks in Fig. 4.17 indicate the deuterated positions of the aromatic anions. A plot of the hydrogen exchange activity of each EDA complex against the sum of the reduction potential ($-E_o$, *V*) and half the localization energy (L_π, eV) of the aromatic hydrocarbons, showed a fairly good linear relationship (Fig. 4.17). The quantity ($-E_o+1/2L_\pi$) is considered to represent the "one-electron localization energy" of each aromatic anion. Thus, heterolytic hydrogen activation by EDA complexes can be understood in terms of the structural properties of their components or of the complexes.

The proposed mechanism of hydrogen activation by polar centers of EDA complexes is shown on page 118:

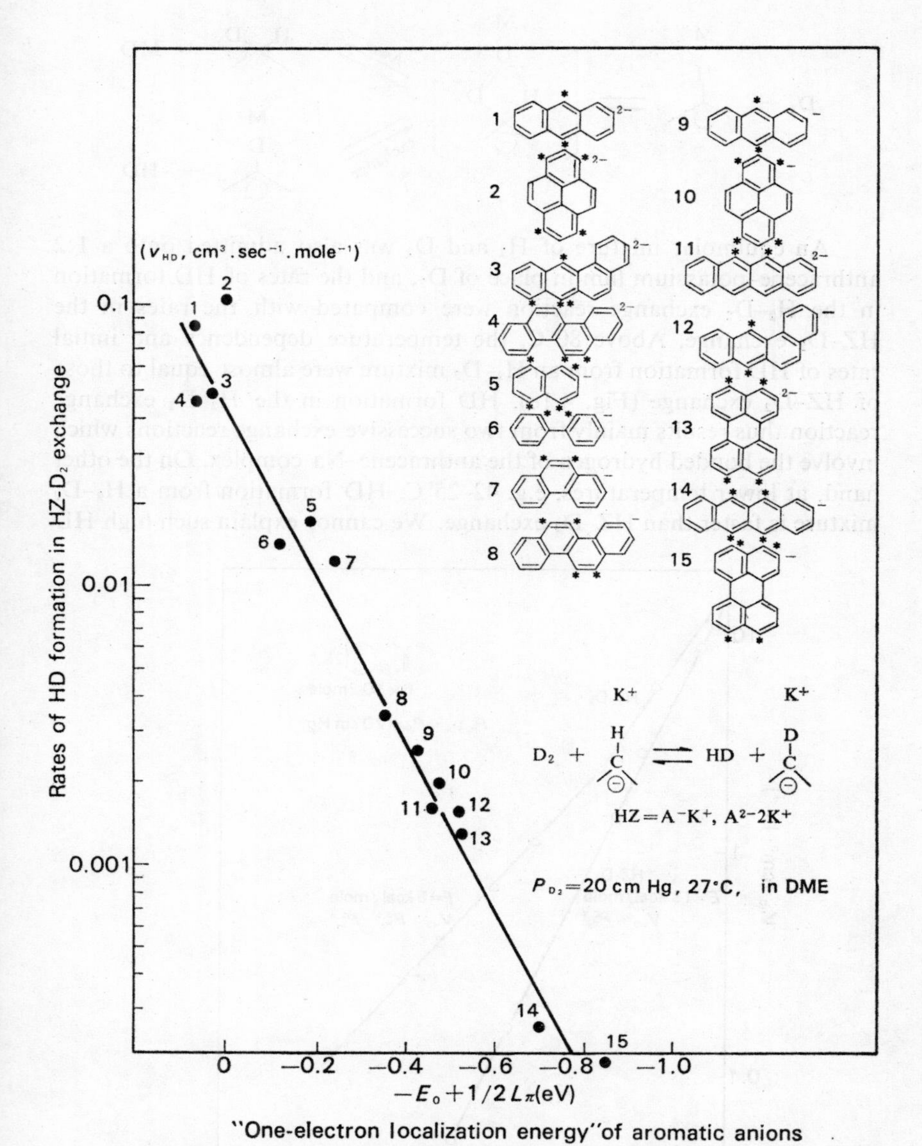

Fig. 4.17 Relative activity of HZ–D_2 exchange of 1:1 and 1:2 EDA complexes of aromatic hydrocarbons with potassium, and electronic properties of aromatic mono- and dianions.

M^+ / H / ⊖ + D_2 ⇌ [M^+ / D, H, D / ⊖] — Ads. ⇌ M^+ / H D / ⊖ + MD; — exchange ⇌ M^+ / D / ⊖ + HD

An equimolar mixture of H_2 and D_2 was also admitted onto a 1:2 anthracene–potassium film in place of D_2, and the rates of HD formation in the H_2–D_2 exchange reaction were compared with the rates in the HZ–D_2 exchange. Above 80°C, the temperature dependency and initial rates of HD formation from an H_2–D_2 mixture were almost equal to those of HZ–D_2 exchange (Fig. 4.18). HD formation in the H_2–D_2 exchange reaction thus results mainly from two successive exchange reactions which involve the bonded hydrogen of the anthracene–Na complex. On the other hand, at lower temperatures, e.g. 42–25°C, HD formation from a H_2–D_2 mixture is faster than HZ–D_2 exchange. We cannot explain such high HD

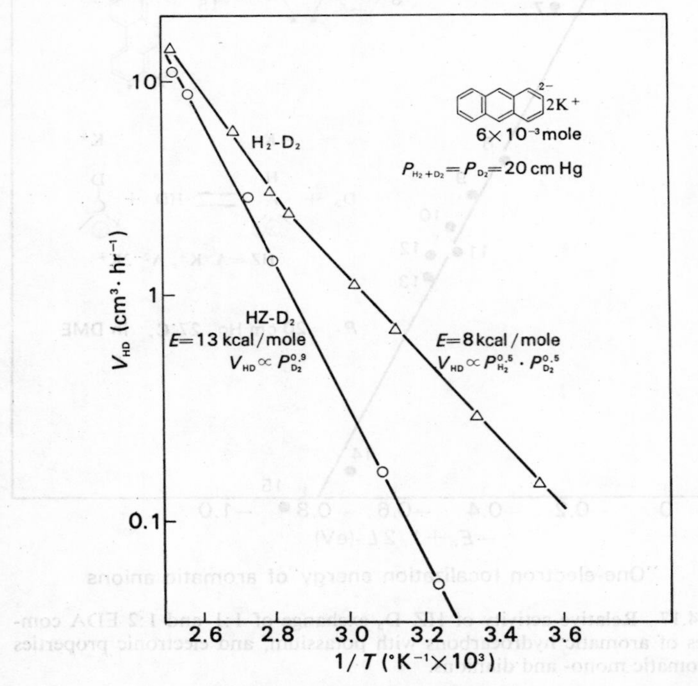

Fig. 4.18 H_2–D_2 and HZ–D_2 exchange reactions over anthracene^{2-}2K$^+$ film.

formation activity by the above mechanism alone. Moreover, the rates of H_2–D_2 exchange reaction are proportional to the square-root of the H_2 and D_2 pressures. Thus, at lower temperatures, the H_2–D_2 exchange reaction may occur mainly through another mechanism, such as recombination of weakly, dissociatively chemisorbed hydrogen on the complex film, i.e. a homolytic chemisorption mechanism.[39,40]

In solutions of these EDA complexes, the H_2–D_2 exchange reaction from a 1:1 mixture of H_2 and D_2 proceeded exclusively through the "bond-exchange mechanism". The rates of exchange and activation energy for H_2–D_2 exchange were almost identical with those for D_2–HZ exchange in anthracene–K (1:2) complex solution[41]

$$\begin{cases} D_2 + HZ \rightleftharpoons HD + DZ \\ H_2 + DZ \rightleftharpoons HD + HZ, \end{cases} \tag{4.7}$$

where HZ denotes the EDA complex.

When tetracene was employed to prepare a 1:2 EDA complex film with cesium, the H_2–D_2 exchange reaction proceeded much faster than the HZ–D_2 exchange over the complex film at all temperatures, as shown in Fig. 4.19.

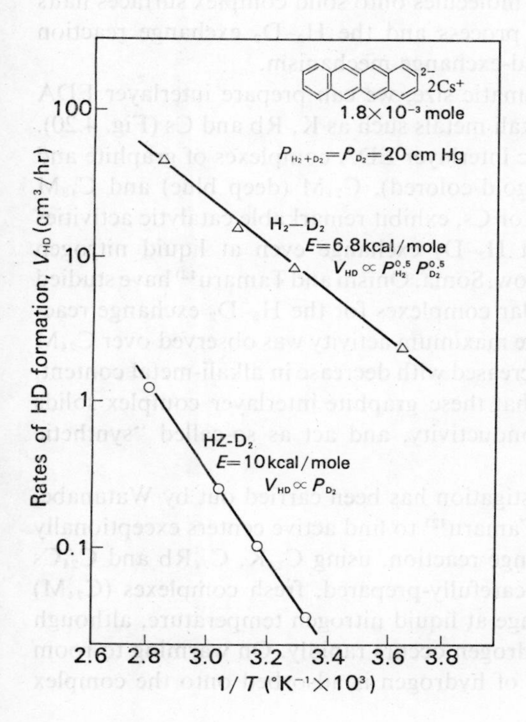

Fig. 4.19 H_2–D_2 and HZ-D_2 exchange reactions over tetracene^{2-}2Cs$^+$ film.

The square-root pressure dependency and smaller activation energy (6.7 kcal/mole) of the H_2–D_2 exchange reaction indicate that it takes place preferentially through homolytic chemisorption over the tetracene–Cs complex film.

$$\left.\begin{array}{l} H_2(g) \rightleftharpoons 2H(a) \\ D_2(g) \rightleftharpoons 2D(a) \end{array}\right\} \rightleftharpoons 2HD(a) \rightleftharpoons 2HD(g) \qquad (4.8)$$

On increasing the size of the aromatic moiety from anthracene to pentacene, homolytic hydrogen activation becomes predominant over the corresponding 1:2 EDA complex films with alkali metals. The H_2–D_2 exchange reaction occurs readily over EDA complex films with larger aromatic components, even at lower temperatures, as it does over typical metal or metal oxide surfaces.

Moreover, the active centers for homolytic chemisorption over solid EDA complexes are particularly sensitive to small amounts of accepting molecules, such as CO, CO_2, O_2, N_2O and NO, which cause partial (CO and CO_2) or complete (O_2, N_2O and NO) inhibition of the H_2–D_2 exchange reaction.[40] They are also sensitive to polar reagents such as THF, DME and polyethers, which probably solvate aromatic anions and alkali metal cations. The addition of polar molecules onto solid complex surfaces halts the homolytic chemisorption process and the H_2–D_2 exchange reaction takes place mainly by the bond-exchange mechanism.

As a limiting case of aromatic size, we can prepare interlayer EDA complexes of graphite with alkali metals such as K, Rb and Cs (Fig. 4.20). It is known that stoichiometric interlayer EDA complexes of graphite and alkali metals, such as C_8M (gold-colored), $C_{24}M$ (deep blue) and $C_{48}M$ (dark blue), where M = K, Rb or Cs, exhibit remarkable catalytic activities for hydrogen adsorption and H_2–D_2 exchange even at liquid nitrogen temperature. Watanabe, Kondow, Soma, Onishi and Tamaru[42] have studied the activities of various lamellar complexes for the H_2–D_2 exchange reaction. As shown in Table 4.6, the maximum activity was observed over $C_{24}M$ complexes, and the activity decreased with decrease in alkali-metal content. Ubbelohde[43] have reported that these graphite interlayer complex solids have remarkable electrical conductivity, and act as so-called "synthetic metals".

Recently, a detailed investigation has been carried out by Watanabe, Kondow, Soma, Onishi and Tamaru[42] to find active centers exceptionally effective for the H_2–D_2 exchange reaction, using $C_{24}K$, $C_{24}Rb$ and $C_{24}Cs$ complexes. They found that carefully-prepared, fresh complexes ($C_{24}M$) are inactive for H_2–D_2 exchange at liquid nitrogen temperature, although the physical adsorption of hydrogen occurs rapidly. On warming to room temperature, a small amount of hydrogen is adsorbed onto the complex

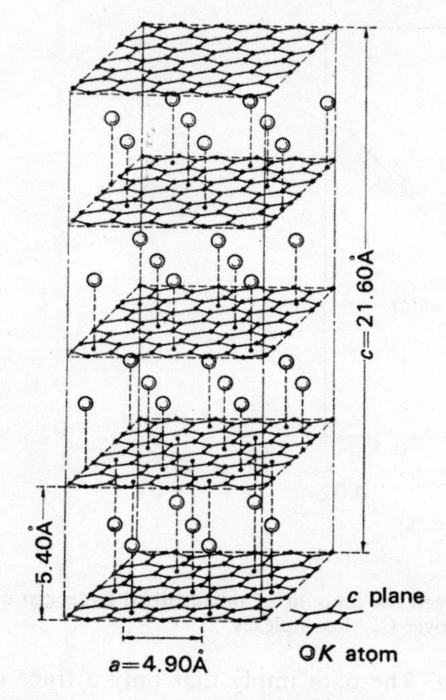

Fig. 4.20 Structural profile of the interlayer EDA complex, C_8K.

solids, and the H_2–D_2 exchange then proceeds rapidly even at liquid nitrogen temperature.

Fig. 4.21 shows the relation between the amount of preadsorbed hydrogen and the H_2–D_2 exchange activity over the $C_{24}K$ lamellar complex

TABLE 4.6 H_2–D_2 exchange reaction over C_8M, $C_{24}M$, $C_{36}M$, $C_{48}M$ and $C_{60-72}M$ complexes of graphite and alkali metals

	H_2, D_2 adsorption (at saturation) ml (STP) / g (graphite)	relative ratio	Catalytic activity in H_2–D_2 exchange: k_1 (hr^{-1}, 0°C)†[1]	volume ratio	E_a†[2] (kcal/mole)
$C_{24}K$	161	100	3.60	100	7.5
$C_{36}K$	115	70	3.06	86	7.1
$C_{48}K$	80	49	1.20	33	6.8
$C_{60-72}K$	61	37	1.02	29	6.0
C_8K	<1	<1	<0.03	<1	—
C_8K hydride†[3] ($C_8K\,H_{2/3}$)	24	15	<0.51	14	6.0

†[1] first-order rate constants for H_2–D_2 exchange (unit per gram of graphite).
†[2] Activation energy.
†[3] The hydride complex prepared at saturated H_2 adsorption by C_8K at 70–80°C (composition, $C_8KH_{2/3}$).

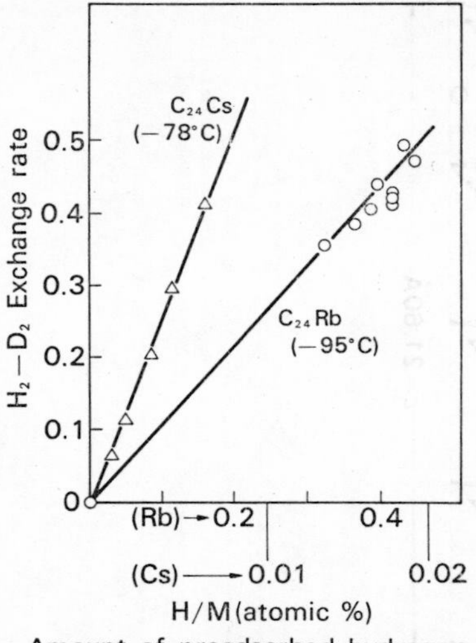

Fig. 4.21 Relation between the amount of preadsorbed hydrogen and the H_2–D_2 exchange activity over $C_{24}M$ complexes.

at liquid nitrogen temperature. The data imply that only a trace of the preadsorbed hydrogen (estimated to be about 0.1% of the total adsorbed hydrogen) is effective in providing active centers for H_2–D_2 exchange. The H_2–D_2 exchange reaction should proceed via the Eley-Rideal mechanism, i.e. a direct exchange between active adsorbed hydrogen and hydrogen molecules, as follows:

$$H_2 + 2\,* \rightleftharpoons 2\underset{*}{\overset{|}{H}}$$

$$\underset{*}{\overset{H}{|}} + D_2 \rightleftharpoons H \cdots D \cdots D \rightleftharpoons HD + \underset{*}{\overset{|}{D}}, \quad (4.9)$$

where * denotes a specific active center on the graphite–alkali metal complex. It seems probable that localized bond-splitting at the edges and/or graphite net planes, which might be generated thermally or by electron donation in complexation, may be responsible for creating the active sites for H_2–D_2 exchange.

In the case of aromatic quinone–alkali metal EDA complexes, the H_2–D_2 exchange reaction proceeds via the "chemisorption mechanism" in the solid state. On increasing the size of the aromatic quinone molecules to incorporate large π-conjugated systems, the rates of the H_2–D_2 exchange

TABLE 4.7 The activities of aromatic quinone–K EDA complexes for H_2—D_2 exchange and the ethylene hydrogenation

Complex	$H_2 + D_2 = 2HD$ ν_{HD} (26°C) (cm³·min⁻¹·mole⁻¹)	E (kcal/mole)	$C_2H_4 + H_2 \rightleftharpoons C_2H_6$ $\nu_{C_2H_6}$ (120°C) (cm³·min⁻¹·mole⁻¹)	E (kcal/mole)
Naphthoquinone⁻ K^+	Inactive		Inactive	
Anthraquinone²⁻ $2K^+$	390	5.3	221	9.3
(pyrene-type quinone) K	875	7.3	120	—
(larger polycyclic quinone) K	994	2.0	10.0	12

reaction increased in EDA complexes with alkali metals, as shown in Table 4.7.[44] For smaller aromatic quinones such as benzoquinone and naphthoquinone, irreversible reduction of the quinones took place, yielding hydroquinone and hydride. The H_2–D_2 exchange reaction over these complexes was negligible.[45]

$$\text{benzoquinone} + 2Na \rightleftharpoons [\text{O}\cdot\text{–C}_6\text{H}_4\text{–O}\cdot]^{2-}\,2Na^+ \xrightarrow{2H_2} \text{hydroquinone} + 2NaH$$

Catalytic activation of hydrogen may be expected over partially delocalized quinone radicals with large π-conjugated redox systems.

Stoichiometric EDA complexes of organometallic compounds such as metal phthalocyanines act as catalysts for hydrogen activation (Fig. 4.22). In forming EDA complexes of phthalocyanines with alkali metals, the metal and metal-free phthalocyanines are reduced to four distinct anions (five for CoPc). Taube[46] and other workers have determined the electronic states of these polyanion complexes from their visible spectra, magnetic susceptibility and esr spectrometry. These complexes are good systems for studying the correlation between the electronic structure of

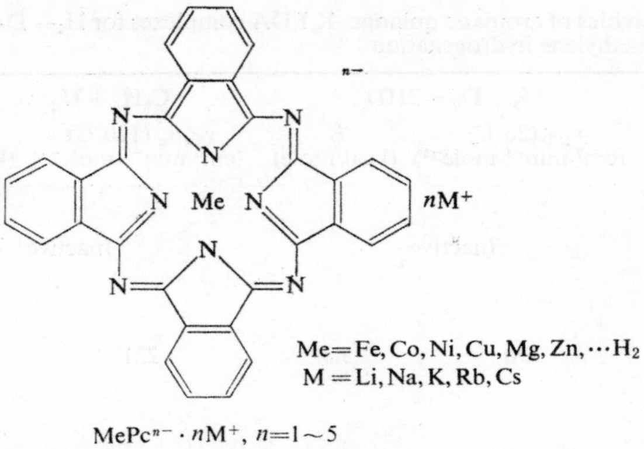

Me = Fe, Co, Ni, Cu, Mg, Zn, ⋯ H_2
M = Li, Na, K, Rb, Cs

$MePc^{n-} \cdot nM^+$, $n=1\sim5$

Fig. 4.22 Phthalocyanine EDA complexes ($MePc^{n-}\ nM^+$).

EDA complexes and their catalytic activity, because of their well-defined electronic states. Moreover, the metal phthalocyanines can be considered as model compounds of biological importance, since they are similar in structure to hemin, chlorophyll and vitamin B_{12}. In particular, hemin and chlorophyll in their working states are sometimes discussed in terms of charge transfer or EDA complex formation.

Naito, Ichikawa and Tamaru[47)] have found that the HZ–D_2 exchange reaction takes place reversibly over various stoichiometric EDA complexes of phthalocyanines with alkali metals. The total number of exchanged hydrogens in the complexes, and the infrared spectra of the deuterated samples, indicate that all the peripheral hydrogen atoms of the benzene rings of the phthalocyanines are exchanged. The rates and activation energy of the HZ–D_2 exchange reaction over various complex films are given in Table 4.8. As the reduction level falls, the rate of hydrogen exchange falls by one order of magnitude and the activation energy increases by about 0.5 kcal/mole. The reactivity does not change when sodium is replaced by lithium, potassium or rubidium. This indicates that the activity for the HZ–D_2 exchange reaction is determined only by the number of electrons ascribable to the π systems of the phthalocyanines. Table 4.8 also shows the correlation between the electronic configuration of polynegative phthalocyanines and the HZ–D_2 exchange activity of the complexes. It can be seen for example that $NiPc^{4-}\ 4Na^+$ and $CoPc^{5-}\ 5Na^+$ complexes, whose electronic configurations are $d^8+\pi^4$, have similar rates and similar activation energies for HZ–D_2 exchange. The circumstances are the same with $NiPc^{3-}3Na^+$ and $CoPc^{4-}4Na^+$ complexes, whose electron configurations are $d^8+\pi^3$. It can be concluded therefore that the HZ–D_2 exchange reaction does not depend upon the central metal atom or the counter cation, but that the reactivity is determined by the number of

TABLE 4.8 Electron configuration of phthalocyanine polyanions and HZ-D_2 exchange over stoichiometric EDA complexes of phthalocyanines with alkali metals

Complexes	Electron configuration $MePc^{n-}$	HZ-D_2 exchange† ν_{HD} ($cm^3 \cdot min^{-1} \cdot g^{-1}$)	(kcal/mole)
$NiPc^{4-}4Na^+$	$d^8+\pi^4$	9.72	11.5
$NiPc^{3-}3Na^+$	$d^8+\pi^3$	1.42	11.9
$NiPc^{2-}2Na^+$	$d^8+\pi^2$	0.201	—
$NiPc^{-}\ Na^+$	$d^8+\pi^1$	0.027	—
$NiPc^{4-}4Li^+$	$d^8+\pi^4$	11.8	11.4
$NiPc^{4-}4K^+$	$d^8+\pi^4$	8.78	11.5
$NiPc^{4-}4Rb^+$	$d^8+\pi^4$	7.65	11.5
$CoPc^{5-}5Na^+$	$d^8+\pi^4$	11.1	11.6
$CoPc^{4-}4Na^+$	$d^8+\pi^3$	1.03	12.0
$CoPc^{3-}3Na^+$	$d^8+\pi^2$	0.136	—
$FePc^{4-}4Na^+$	$d^8+\pi^2$	0.159	12.2

† $P_{D_2}=27$ cm Hg, 200°C.

electrons ascribable to the π-conjugate systems of the phthalocyanines, as demonstrated in Fig. 4.23.

Naito, Ichikawa, Soma, Onishi and Tamaru[47,48] have found that H_2–D_2 exchange occurs much faster than HZ–D_2 exchange through the chemisorption mechanism, as is the case with the tetracene^{2-}2Cs^+ film mentioned above. Fig. 4.24 shows the activities for H_2–D_2 exchange with various central metal atoms, and the degree of reduction of the phthalo-

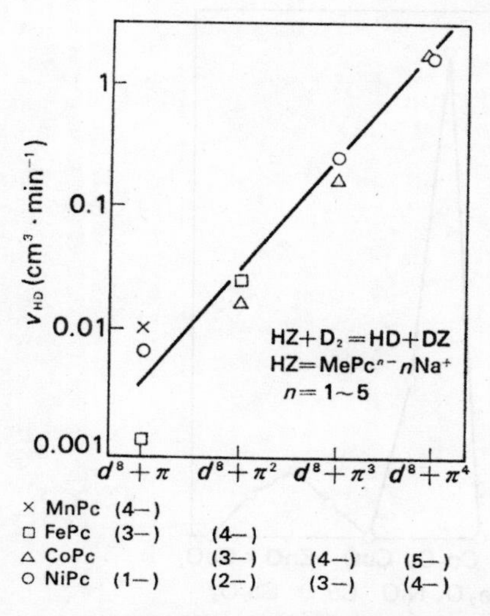

Fig. 4.23 HZ-D_2 exchange reaction over $MePc^{n-}nNa^+$ and electron configuration of phthalocyanine polyanions (P_{D_2}=27 cm Hg; 200°C).

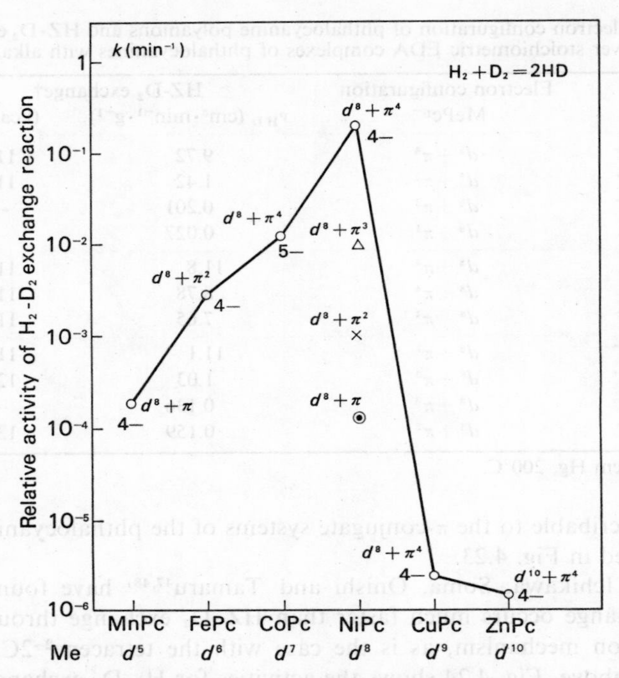

Fig. 4.24 H_2–D_2 exchange reaction over $MePc^{n-}nNa^+$ films and electron configuration of phthalocyanine polyanions ($P_{H_2}=P_{D_2}=13.5$ cm Hg; 60°C).

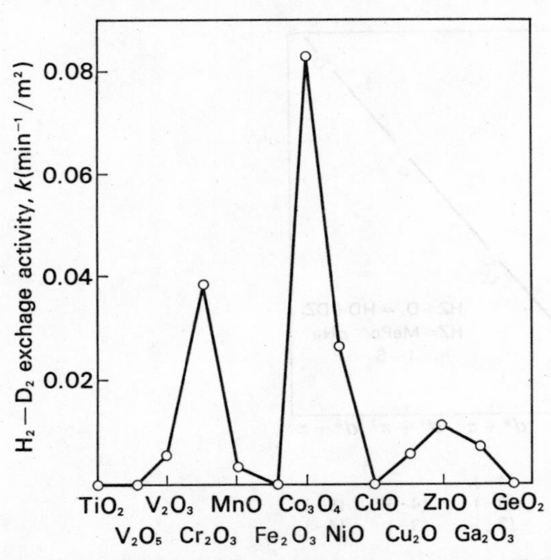

Fig. 4.25 Variation of activity of metal oxides of the fourth period in the H_2–D_2 exchange reaction.

cyanines. The reactivity of the complexes for H_2–D_2 exchange was markedly dependent upon the central metal atom in the order, Ni > Co > Fe > H_2 > Mn > Mg > Cu > Zn. The rate fell one order of magnitude for each decrease of the reduction state. The order of activity for this chemisorption process (Ni > Co > Fe > Mn) is thus in good accord with the usual crystal field stabilization energies of divalent transition metal cations. Moreover, the reactivity of the Mg, Cu and Zn complexes, which have little or no crystal field stabilization energy, is almost identical. Dowden *et al.*[49] have obtained a similar correlation for the H_2–D_2 exchange activity over transition metal oxides in relation to the crystal field energy of the respective metals (*cf.* Fig. 4.25).

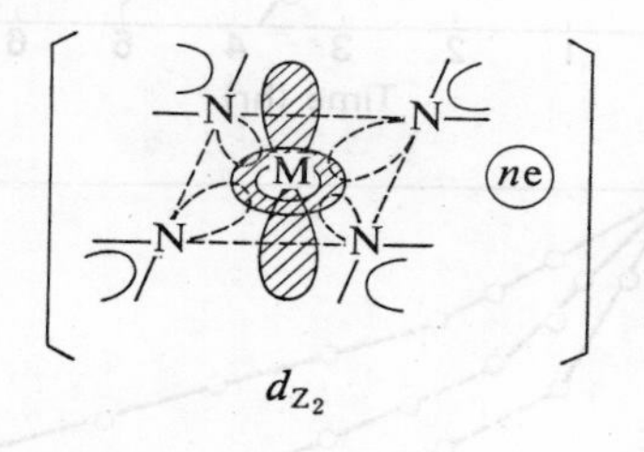

4.2.3 Hydrogen exchange reaction of acetylene

Ichikawa, Soma, Onishi and Tamaru[50] have investigated the kinetics of acetylene exchange over various EDA complex films, and discussed the correlation between the exchange activities and electronic structures of the complexes.

When C_2D_2 was admitted over evaporated films of metal-free, magnesium-, iron-, cobalt-, sodium- and copper-phthalocyanines, no hydrogen exchange could be detected at 200°C, even under ultraviolet illumination. However, after exposure to sodium vapor at room temperature, the films exhibited a high activity for the exchange reaction of hydrogen with acetylene (C_2D_2), and readily formed C_2HD at room temperature according to the following equations:

$$C_2D_2 + HZ = C_2HD + DZ \tag{4.10}$$

$$C_2H_2 + DZ = C_2HD + HZ, \tag{4.11}$$

where HZ represents the phthalocyanine EDA complexes.

On exposure to sodium vapor, no gas evolution was detected and the phthalocyanine films changed color to some extent; Fe- and Co-Pc to a faint red-violet, and Cu- and Mg-Pc to blue-violet. The rate of hydrogen

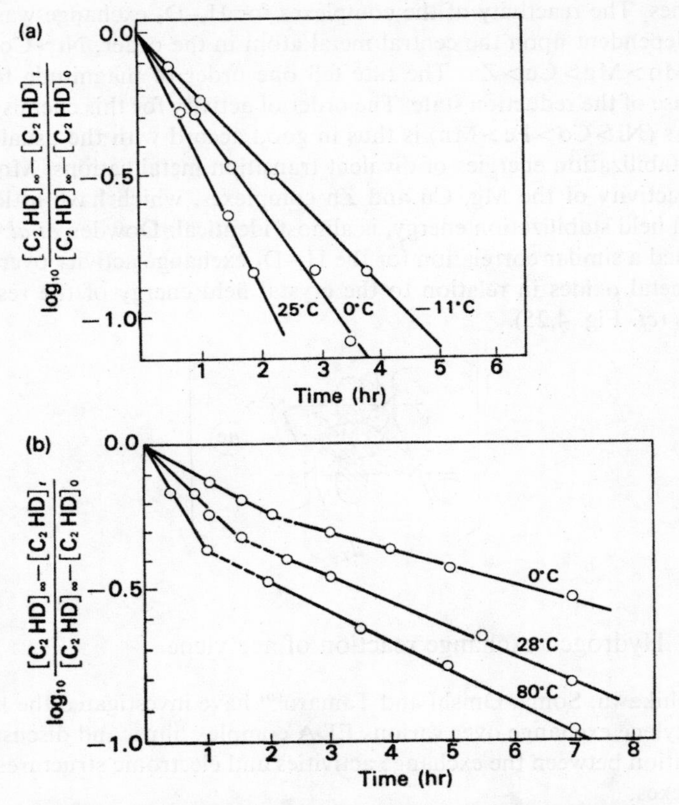

Fig. 4.26 (a) Exchange reaction of hydrogen between acetylene and Fe-phthalocyanine–Na complex. (b) Exchange reaction of hydrogen between acetylene and metal-free phthalocyanine–Na complex.

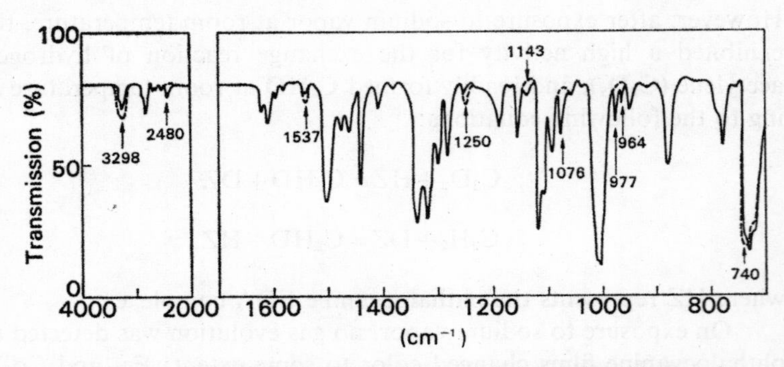

Fig. 4.27 Infrared spectrum of metal-free phthalocyanine–Na complex. ---, Original complex film; —, after reaction with C_2D_2 at 25°C.

exchange was studied under 15 cm Hg acetylene pressure at constant volume (320 ml) in the temperature range, $-10°$ to 80°C. The number of hydrogen atoms located in acetylene molecules was substantially larger than that in the surface of the complexes and, accordingly, in the gas phase, the concentration of C_2D_2 was much larger than that of C_2HD, while that of C_2H_2 was very low throughout the reaction. Consequently, the appearance of C_2HD obeyed first-order kinetics, as shown in Fig. 4.26(a), the rate constant being independent of the number of exchangeable hydrogens on the surface. The first-order kinetics imply uniform reactivity of the exchangeable hydrogens in the surface. The rate is proportional to the acetylene pressure.

The rate curve for the metal-free phthalocyanine EDA complex with sodium (H_2Pc-Na) seems to consist of two different first-order curves, which have different slopes in the initial period and in the later stage of the reaction, as shown in Fig. 4.26(b). The ratio of the numbers of the two kinds of sites was approximately four, which may be understood by taking two kinds of hydrogen atoms into consideration: those at the N–H bond in the center of the molecule and those in the outermost peripheral regions (two in each of the surrounding benzene rings).

The position of the exchangeable hydrogen in phthalocyanine complexes was studied using an infrared technique. The cell had two NaCl windows, and NaCl or KBr plate, and two side-arms. A small amount of phthalocyanine was first sublimed onto the NaCl (or KBr) plate from one of the side-arms and then sodium vapor was brought into contact with the phthalocyanine film from the other arm. The phthalocyanine film remained in the β-form throughout the experiments. When deuteroacetylene was introduced into the cell, the NH absorption peaks of the H_2-Pc complex decreased and new peaks gradually appeared, as shown in Fig. 4.27 (the solid line). The new absorption peaks were assigned to N–D bonds in the center of the molecule, since the peaks were in exact agreement with those of D_2-Pc obtained by washing Na-Pc with D_2O. These results thus show that the hydrogen atoms in the center of metal-free phthalocyanines can be exchanged with those in acetylene, when an EDA complex is formed with sodium.

As to metal-phthalocyanines such as Mg- and Cu-Pc, no appreciable change was detected upon exposure to C_2D_2 for a weak band at about 540 cm^{-1}. This band disappeared on treatment of the film with C_2H_2. It is thus possible to assign this band to the C–D out-of-plane vibration of the carbon-hydrogen bonds in the peripheral benzene rings of the molecules.

Chlorpromazine, a well-known tranquilizer, is a strong electron donor. The reactivities of the EDA complexes with various organic electron acceptors have been studied.[51] It was shown that the C_2D_2–HZ exchange reaction proceeded reversibly over EDA complexes of chlorpromazine at a considerable rate at room temperature, although no such reactivity was observed over chlorpromazine or the acceptors alone under the same reaction conditions. It was therefore suggested that chlorpromazine may form "active" EDA complexes with many organic electron acceptors, especially quinones, and that this may perhaps be correlated with the pharmacological activity of the tranquilizers (see Chapter 5).

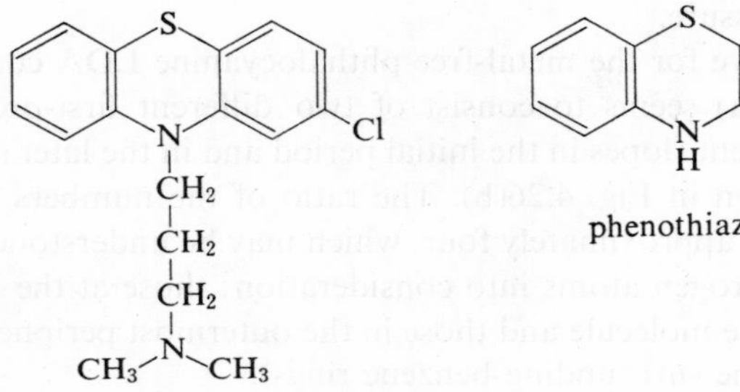

chlorpromazine

phenothiazine

Evaporated films (0.001 mole) of chlorpromazine and phenothiazine were prepared on the wall of a glass vessel, onto which small amounts of acceptors (0.002 mole) such as 2,3-dicyano-, *p*-benzo-, α-naphtho-, 2,3-dichloro-5,6-dicyano-(DDQ)quinones, and also *p*-chloranil, were sublimed *in vacuo*. As a result of this contact, deep-colored complexes were obtained. Similar EDA complexes were prepared by mixing solutions of chlorpromazine and each acceptor either *in vacuo* or in an atmosphere of nitrogen, and evaporating the solvent. When C_2D_2 was introduced over the complexes at room temperature, a considerable amount of C_2HD was formed as a result of hydrogen exchange. The reaction took place reversibly. The reaction was first order with respect to acetylene and also with respect to the amount of hydrogen to be exchanged in the complexes, which was estimated from the composition and pressure (15 cm Hg) of the isotopic mixture of acetylene in the temperature range between 0 and 120°C. For EDA complexes of chlorpromazine and *N*,*N*,*N*′,*N*′-tetramethyl-*p*-phenylenediamine with various acceptor molecules, the first-order rate constants for the exchange reaction are shown in Table 4.9. The activation energies so far obtained for each rate constant lie between 7 and 11 kcal/mole.

In a similar manner, the exchange kinetics over films prepared by

TABLE 4.9 C_2D_2–HZ exchange reaction over organic EDA complexes†

Donor / Acceptor	EA (eV)	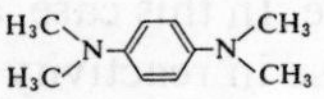 k (90°C)	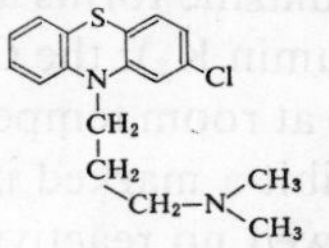 k (50°C)
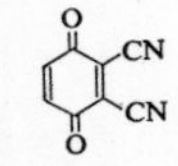	1.7	+++	+++
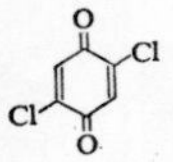	1.2	++	++
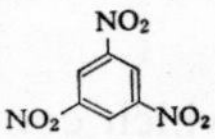	0.8	+	+
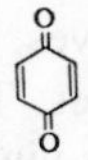	0.7	++	++
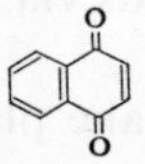	0.6	++	++
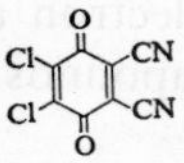	1.9	—	—
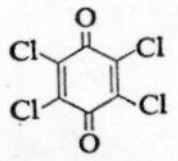	1.37		—

† k = first-order rate constants; +++ (~10^{-1} hr^{-1}), ++ (10^{-1}~10^{-2} hr^{-1}), +(10^{-2} ~10^{-3} hr^{-1}), and – (no exchange). EA = electron affinity from Briegleb.

mixing benzene solutions of the components were studied. The hydrogen exchange reaction did not proceed over these donor or acceptor molecules alone under otherwise identical reaction conditions.

EDA complexes of chlorpromazine with such electron acceptors as 2, 3-dicyanoquinone were found to be much more reactive for the C_2D_2–HZ exchange reaction than those of phenothiazine with the same corresponding acceptors. As shown in Table 4.10, EDA complexes of chlorpromazine with quinones were much more active than those with trinitrobenzene or pyromellitic acid anhydride, which are stronger electron acceptors. It may be suggested that the >C=O groups of the quinones interact with the two nitrogen atoms of the chlorpromazine molecule to yield "active" EDA complexes. From the biochemical point of view, it is of interest that chlorpromazine forms a stable EDA complex with 2-methyl-α-naphthoquinone (vitamin K_3); the C_2D_2–HZ exchange reaction proceeded at a considerable rate at room temperature. In this case, a tranquilizer and a vitamin K_3 thus exhibit a marked increase in reactivity by forming an EDA complex: they showed no reactivity separately.

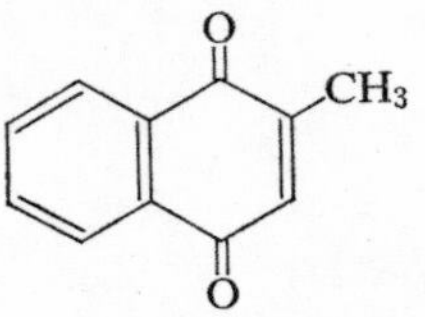

2-methyl-α-naphthaquinone

When *p*-chloranil or 2,3-dichloro-5,6-dicyano-*p*-quinone, which contains no hydrogen and is a comparatively strong electron acceptor, was employed to form EDA complexes with phenothiazine derivatives, the C_2D_2–HZ exchange reaction did not take place over the complexes. This suggests that the hydrogen exchange reaction of acetylene proceeds via the hydrogen in the anion radicals.

Similar hydrogen exchange reactions have been found to take place over EDA complexes of metallocenes with various organic electron acceptors such as quinones, and nitro- and cyano-substituted compounds.[52]

4.2.4 Photocatalysis over EDA complexes

Investigations have also been made by Ichikawa, Soma, Onishi and Tamaru[52] on the photocatalytic hydrogen exchange reaction between acetylene and organic EDA complexes. When C_2D_2 (15 cm Hg) was admitted over EDA complexes of various phthalocyanines with organic electron acceptors such as 2,3-dicyanoquinone at temperatures of between 25 and 90°C, the hydrogen exchange reaction of acetylene proceeded at a negligible rate and the components of the complexes were separated from each other above 100°C as a result of sublimation. Under illumination by a mercury lamp, on the other hand, a considerable amount of C_2HD appeared slowly at 60°C. The photocatalytic hydrogen exchange reaction did not take place over either the phthalocyanines or organic electron acceptors alone. Calvin and his co-workers have demonstrated[8,9] that the photoconductivity of phthalocyanines increases markedly by forming a double layer with organic electron acceptors, and the increase was attributed to charge separation from donor to acceptor. The activity of the

TABLE 4.10 Rate constants for the *ortho/para* hydrogen conversion of anthracene and anthracene–TNB complex

Compounds	k (hr^{-1}) In the dark	Under illumination
Blank (glass surface)	$\sim 1\times 10^{-3}$	$\sim 1\times 10^{-3}$
Anthracene	$\sim 1\times 10^{-3}$	$1.8_3\times 10^{-2}$
Anthracene–TNB	$3.6_2\times 10^{-3}$	$3.7_0\times 10^{-2}$

complexes for the C_2D_2–HZ exchange reaction under illumination may be attributable to a sort of electron transfer, forming anion radicals.

Inokuchi, Tsuda and Kondow[53] have presented results for *o-/p-*hydrogen conversion in the presence of anthracene vapor, and in the presence of anthracene–1,3,5-trinitrobenzene(TNB) CT complex in the gas phase, under illumination. Assuming that the rate of *ortho/para* hydrogen conversion obeys a first-order law, the apparent rate constants (k) may be calculated as in Table 4.10. In the presence of anthracene vapor, the k value for the interconversion in the dark was not significantly different from that in the blank tests; this was also the case for the H_2–D_2 exchange reaction. Under illumination, however, a fairly strong *ortho/para* hydrogen conversion was observed; the rate of conversion is given in the right-hand column of Table 4.10. Slight H_2–D_2 exchange also occurred in this case. In the presence of anthracene–TNB complex, a very slow *ortho/para* conversion took place in the dark, while illumination accelerated the conversion rate strongly. In these experiments, the compounds appeared to acquire a paramagnetic character as a result of irradiation, since *ortho/para* hydrogen conversion took place under illumination and the H_2–D_2 exchange rate was very slow in comparison with that of the interconversion. Such paramagnetism may be considered as being caused by the excited state of the anthracene–TNB complex.

4.3 Selective Hydrogenation of Unsaturated Hydrocarbons over Various EDA Complexes

Ichikawa, Soma, Onishi and Tamaru[54] have reported that a highly selective hydrogenation of olefins and acetylenic compounds is observed over EDA complexes of sodium. The selectivity was attributed to the controlled poisoning of the catalyst surface by the chemisorption of hydrogen and unsaturated compounds. A mixture of ethylene and hydrogen gas (1:1 mixture, 25 cm Hg total pressure) was admitted to EDA complexes of sodium (~0.01 mole) with phthalonitrile, anthracene, *p*-quinone, tetracyanobenzene, *p*-chloranil, pyrene, or violanthrene B, etc. (~0.001 mole) at temperatures between 50 and 120°C. Small amounts of ethane (the yields were estimated to be about 10%) were formed in 40 hr only over EDA complexes of such acceptors as tetracyanobenzene, tetrachlorophthalonitrile, anthraquinone, perylene, and violanthrene B. Propylene (10 cm Hg) was also hydrogenated at 100°C over these EDA complexes, and the yield of propane in 40 hr was of the same order of magnitude as that in ehtylene hydrogenation over each of the complexes. It is interesting to note that the rate of ethylene or propylene hydrogenation was dependent upon the relative partial pressures of the ethylene or propylene to hydrogen.

When a mixture of H_2–D_2 and ethylene (or propylene) was admitted to these EDA complexes, the H_2–D_2 exchange reaction was retarded by increasing the relative content of ethylene (or propylene), which suggests that ethylene (or propylene) and hydrogen are adsorbed competitively over the complexes. The hydrogenation of olefins, accordingly, seems to proceed more easily when the hydrogen and olefins are adsorbed in comparable amounts on the catalyst surface.

Acetylene, a stronger acceptor gas, was not hydrogenated over these EDA complexes, probably because its chemisorption is too strong to allow hydrogen adsorption. The H_2–D_2 exchange reaction proceeded negligibly even at higher temperatures over the active EDA complexes when a small amount of acetylene was mixed with the hydrogen. However, when weaker acceptor gases such as methylacetylene, butadiene, and dimethylacetylene (10 cm Hg) were added to the H_2–D_2 mixture, the H_2–D_2 exchange reaction proceeded over the EDA complexes at 100°C, and hydrogenation took place to yield propylene, a mixture of 2-and 1-butenes, and *cis*-2-butene, respectively. It was also found that the H_2–D_2 exchange reaction was only partially retarded by these olefins or methyl- and dimethylacetylene. The extent of the retardation by such gases is correlated with their affinity with the complex surface. Consequently, it is reasonable that methylacetylene should be hydrogenated only to propylene and not to propane, when methylacetylene was present in the ambient gas. A similar selective hydrogenation was observed in the case of butadiene and dimethylacetylene over these EDA complexes, where only butenes were formed in the first stage and then butane produced subsequently.

When an artificial graphite was employed as an electron acceptor to form a complex with sodium or potassium, the hydrogenation of ethylene, propylene, methylacetylene, 1-butene, and butadiene proceeded at a considerable rate at 25°C under similar reaction conditions to those employed in the other systems. From methylacetylene, propylene and propane were produced quantitatively, step by step with a complete selectivity. Butadiene was rapidly hydrogenated first to a mixture of *cis*-2- and *trans*-2-butenes and then to butane at 25°C over the graphite complexes with sodium or potassium. When a mixture of dimethylacetylene and D_2 was introduced into the graphite–Na complex system pretreated with D_2 at 80°C, *cis*-2-butene-d_2 only was obtained at 25°C in 40 hr.

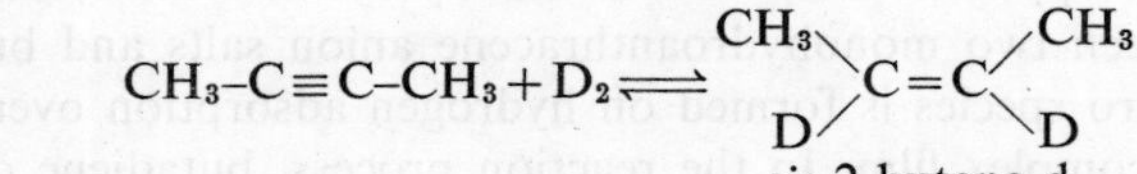

$$CH_3{-}C{\equiv}C{-}CH_3 + D_2 \rightleftharpoons (CH_3)(D)C{=}C(CH_3)(D)$$

cis-2-butene-d_2

$$\text{anthracene}^{2-}\,2Na^+ + D_2 \rightleftharpoons \text{9-H,9-D-anthracene}^{-}\ Na^+ \cdot NaD$$

$$CH_2{=}CH{-}CH{=}CH_2 + \text{9-H,9-D-anthracene}^{-}\ Na^+ \xrightarrow{\text{fast}} \text{anthracene} + C_4H_6D^-Na^+$$

$$C_4H_6D^-Na^+ + \text{9-H,9-D-anthracene}^{-}\ Na^+ \xrightarrow{\text{slow}} \text{anthracene}^{2-}\,2Na^+ + \left\{\begin{array}{l}\textit{cis}\text{-2-butene}\\ \textit{trans}\text{-2-butene}\end{array}\right.$$

(1,4-d_2, 1-d_1)

$C_4H_6D^-Na^+$ $H_2C{\cdots}C(H){\cdots}C(H)(CH_2D)$ Na^+ *syn-* $H_2C{\cdots}C(H){\cdots}C(CH_2D)(H)$ Na^+ *anti-*

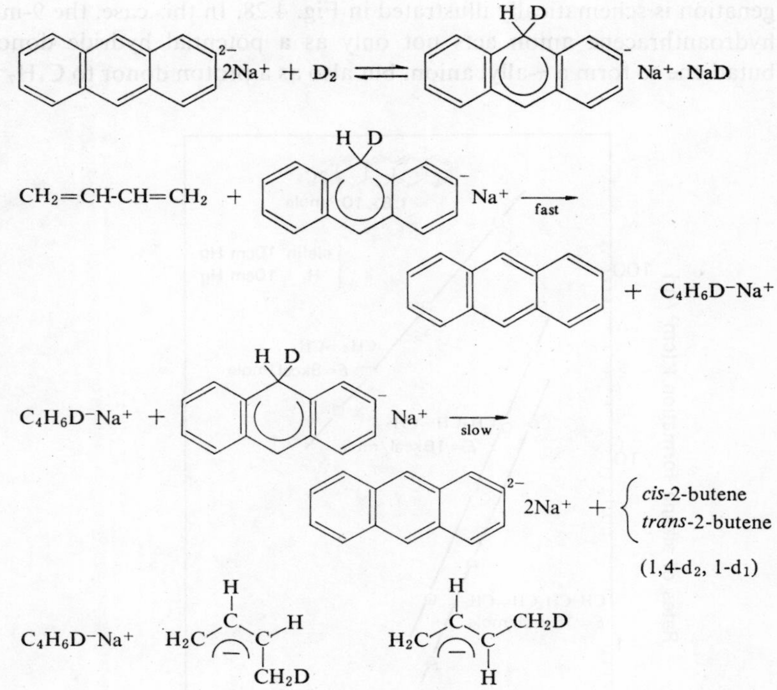

Fig. 4.28 Proposed mechanism for selective hydrogenation of butadiene over anthracene^{2-}2Na^+ film, and the structures of the intermediate π-butenyl carbonion salts.

Ichikawa, Soma, Onishi and Tamaru[55] have found that the hydrogenation of olefins takes place over certain stoichiometric EDA complexes catalyzing dissociative hydrogen chemisorption. For example, butadiene is hydrogenated selectively to an equimolar mixture of *cis*- and *trans*-2-butenes with a small amount of 1-butene over a 1:2 anthracene–Na film. When butadiene is admitted onto a complex film with preadsorbed D_2, *cis*-2-butene-1,4-d_2 and 2-butene-1-d_1 are obtained selectively according to microwave spectroscopic analysis. From the location of the D atoms, and also from the spectral changes during the hydrogenation process, *cis*- and *trans*-2-butenes appear to be produced mainly by the ionic 1,4-additional reaction between two monohydroanthracene anion salts and butadiene. The monohydro species is formed on hydrogen adsorption over the anthracene–Na complex film. In the reaction process, butadiene combines with a hydride ion from the 9-monohydro anion salt to give an allylic intermediate ($C_4H_7^-Na^+$). The proposed mechanism of butadiene hydrogenation is schematically illustrated in Fig. 4.28. In this case, the 9-monohydroanthracene anion acts not only as a potential hydride donor to butadiene to form a π-allyl anion, but also as a proton donor to $C_4H_7^-Na^+$

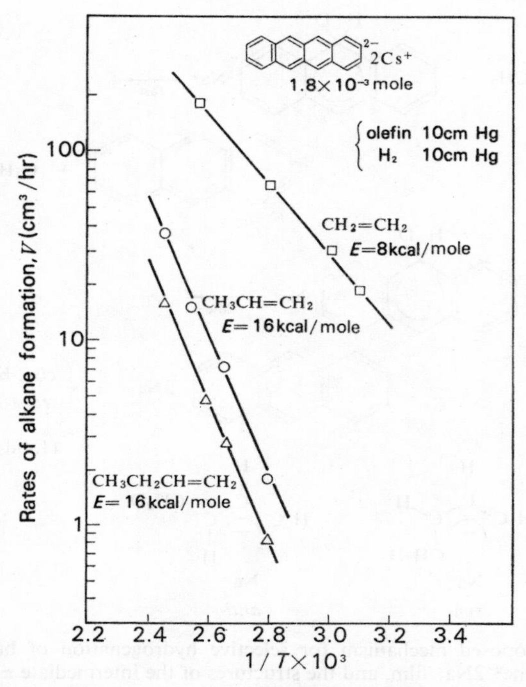

Fig. 4.29 Hydrogenation of substituted olefins over tetracene^{2-}2Cs$^+$ film.

as an intermediate in the formation of an equimolar mixture of *cis*- and *trans*-2-butenes.

Ichikawa and Tamaru have also demonstrated[56] that mono-olefins such as ethylene and propylene are also catalytically hydrogenated over a 1:2 tetracene–Cs complex film. The relative activities for the hydrogenation of several olefins and the activation energies are presented in Fig. 4.29. The results suggest that the electron donating alkyl-substituted olefins are difficult to hydrogenate over this complex film.

It was found[56] that the activities for the hydrogenation of olefins depend markedly upon the alkali metal. As shown in Fig. 4.30, ethylene is

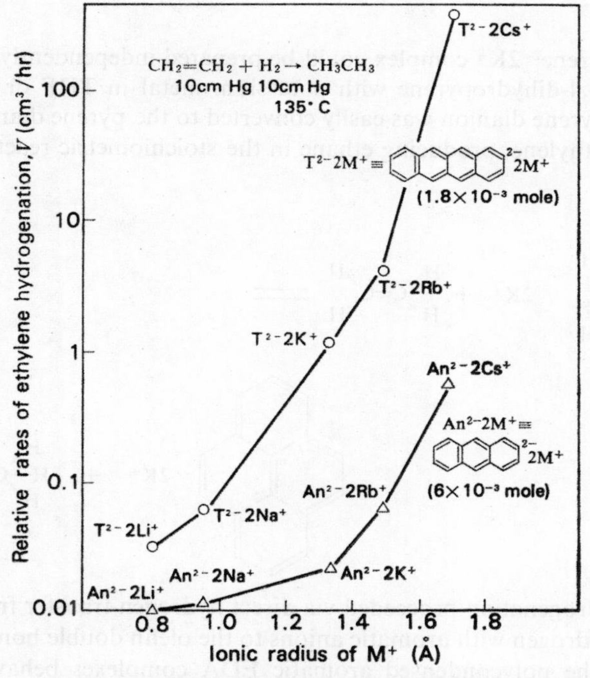

Fig. 4.30 Relative ethylene hydrogenation activity over the 1:2 EDA complexes of tetracene and anthracene with various alkali metals.

readily hydrogenated over 1:2 EDA complex films of higher alkali metals such as cesium, rubidium and potassium. This is attributed mainly to the larger electron-donating complex surface, which facilitates ethylene adsorption. In the EDA complexes of aromatic compounds, the aromatic anions are highly negatively polarized, with the counter cations having a larger ionic radius.

As a model system, the hydrogenation of ethylene was investigated in detail over a pyrene^{2-}2K^{+} complex. Hydrogen adsorption took place reversibly over the complex to yield an adsorbed species, which was elucidated spectroscopically as the 3,4-dihydropyrene dianion complex, as illustrated below:

$$\text{pyrene}^{2-}\ 2K^+ + H_2 \rightleftharpoons \text{3,4-dihydropyrene}^{2-}\ 2K^+$$

3,4-Dihydropyrene^{2-}2K^{+} complex could be prepared independently by the reduction of 3,4-dihydropyrene with potassium metal in THF or DME. This dihydropyrene dianion was easily converted to the pyrene dianion on contact with ethylene, producing ethane in the stoichiometric reaction:

$$\text{3,4-dihydropyrene}^{2-}\ 2K^+ + H_2C{=}CH_2 \rightleftharpoons \text{pyrene}^{2-}\ 2K^+ + H_3C{-}CH_3$$

The olefin hydrogenation proceeded via direct hydrogen transfer from the σ-adsorbed hydrogen with aromatic anions to the olefin double bond. This implies that the polycondensed aromatic EDA complexes behave as a "giant olefin", which, by reversible hydrogenation/dehydrogenation, gives rise to catalytic olefin hydrogenation.

4.3.1 Hydrogen exchange reaction of olefins and isomerization over EDA complexes

It is desirable to determine the reaction intermediates and adsorbed states of olefins during hydrogenation and isomerization over EDA complexes in connection with their structural properties. Microwave spectroscopy may be effective for elucidating the mechanism and reaction intermediates of hydrogen exchange, while double-bond migration of olefins may be studied from the deuterium distribution in the olefin molecules.[57)]

$$C_3H_6 + xD_2 \rightleftharpoons C_3H_{6-x}D_x + xHD \quad (x=1,2,\ldots 6)$$

The product, propylene-d$_1$ (C_3H_5D), in the above case has four geometrical isomers, which are illustrated in Fig. 4.31, and propylene-d$_2$ ($C_3H_4D_2$) has seven isomers. As the dipole moments and moments of inertia of the various deuterated propylene species are different, microwave spectrocopy

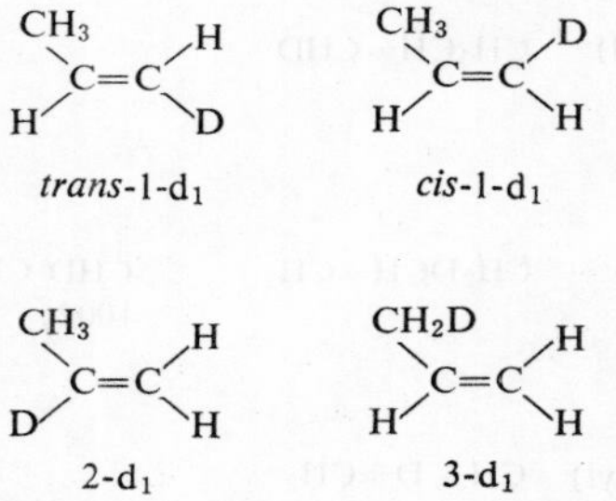

$(CH_3)(H)C{=}C(H)(D)$ — *trans*-1-d$_1$

$(CH_3)(H)C{=}C(D)(H)$ — *cis*-1-d$_1$

$(CH_3)(D)C{=}C(H)(H)$ — 2-d$_1$

$(CH_2D)(H)C{=}C(H)(H)$ — 3-d$_1$

Fig. 4.31 Isomers of propylene-d$_1$.

can not only distinguish between propylene-d$_1$ and -d$_2$ but can also determine the amounts of each of the geometrical isomers of propylene-d$_1$ and -d$_2$, even in a mixture of all possible isomers.

In the hydrogen exchange reaction between propylene and deuterium, many possible mechanisms can be considered as follows: those with *n*-propyl ($CH_3\text{–}CH_2\text{–}CH_2*$), *iso*-propyl ($CH_3\text{–}\underset{*}{C}H\text{–}CH_3$), *n*-propenyl ($CH_3\text{–}CH{=}CH*$), *iso*-propenyl ($CH_3\text{–}\underset{*}{C}{=}CH_2$), σ-allyl ($\underset{*}{C}H_2\text{–}CH{=}CH_2$), and π-allyl ($CH_2 \overset{...}{=} CH \overset{...}{=} CH_2$) intermediates, as well the concerted (push-pull) mechanism. As shown in Fig. 4.32, the characteristic patterns of deuterium distribution of the products, propylene-d$_1$ and -d$_2$, correspond to the different distinct reaction intermediates of the propylene-deuterium exchange.

Kondo, Ichikawa, Saito and Tamaru[58)] have studied the propylene-deuterium exchange reaction over graphite-interlayer compounds such as $C_{24}K$. When a mixture of propylene and D_2 was introduced over the $C_{24}K$ complex at 120°C, simultaneous hydrogenation and hydrogen exchange of the propylene took place. The deuterium distribution of propylene-d$_1$ and

	Reaction intermediates		Product propylene-d_1	Product propylene-d_2
1.	CH_3–CHD–CH_2 (surface-bound)	(*n*-propyl)	$CH_3CD{=}CH_2$	
2.	CH_3, H, CH_2D on C (surface-bound)	(*iso*-propyl)	$CH_2DCH{=}CH_2$ 60% $CH_3CH{=}CHD$ 40%	
3.	CH_3-$CH{=}CH_2$ with H and D (surface)	(push-pull)	$CH_2DCH{=}CH_2$	$CH_2DCH{=}CHD$ 100%
4.	CH_3–CH=CH; D, H (surface)	(*n*-propenyl)	$CH_3CH{=}CHD$	
5.	CH_2=CH–CH_2; D, H (surface)	(σ-allyl)	$CH_2DCH{=}CH_2$	$CHD_2CH{=}CH_2$ 100%
6.	CH_3, CH_2 on C; D, H (surface)	(*iso*-propenyl)	$CH_3CD{=}CH_2$	
7.	CH_2···CH···CH_2; D, H (surface)	(π-allyl)	$CH_2DCH{=}CH_2$	$CHD_2CH{=}CH_2$ 50% $CH_2DCH{=}CHD$ 50%

Fig. 4.32 Possible reaction intermediates and propylene-d_1 and -d_2 initially formed in deuterium-propylene exchange.

-d_2 was observed by microwave spectroscopic analysis as shown in Fig. 4.33, and it was found that propylene-3-d_1 is the only reaction product in the initial stage of the exchange. No propylene-2-d_1 was observed. This indicates that *n*- and *iso*-propyl species cannot be intermediates of this exchange reaction. From the deuterium distribution of propylene-d_2 (propylene-3,3-d_2 50%, 1,3-d_2 47%) at the initial stage, it was suggested that the 1 and 3 carbons are equivalent in the reaction intermediate. The initial distributions of propylene-d_1 and -d_2 can be explained by a π-allyl intermediate alone. It was thus concluded that propylene-deuterium exchange occurs most probably through a π-allyl intermediate over the $C_{24}K$ complex. By similar analytical processes, it was found that propylene-

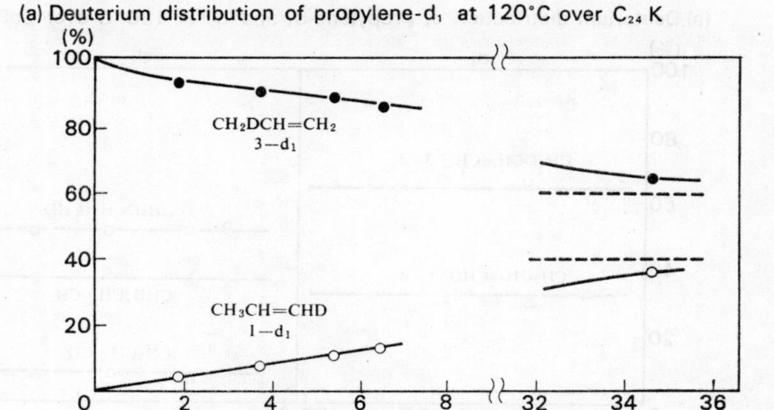

(b) Deuterium distribution of propylene-d_2 ($\Phi=6.6\%$)

1,1-d_2	3%
1,3-d_2	47%
3,3-d_2	50%

(c) Dissociative π-allyl intermediate over $C_{24}K$

$$CH_3CH{=}CH_2 \xrightarrow{-H} [CH_2 \cdots CH \cdots CH_2]$$

$$\downarrow +D$$

$$d_1 \quad CH_2DCH{=}CH_2 + CH_3CH{=}CHD$$

$$\downarrow -H\ +D$$

$$d_2 \begin{cases} CH_2DCH{=}CHD\ 50\% \\ CHD_2CH{=}CH_2\ 50\% \end{cases}$$

Fig. 4.33 The D_2-propylene exchange reaction over $C_{24}K$, and proposed mechanism for deuterium incorporation through a π-allyl intermediate.

deuterium exchange can also take place via π-allyl species (as an adsorbed state) over 1:2 anthraquinone–K and 1:2 tetracene–Cs EDA complexes, which implies that the EDA complexes of aromatic compounds with alkali metals act as typical base catalysts, like ZnO and MgO.

Naito, Ichikawa and Tamaru[59] have observed the deuterium-propylene exchange reaction over various stoichiometric EDA complexes of metal phthalocyanines with sodium. By microwave spectrometry the deuterium distribution of propylene-d_1 and -d_2 was analyzed in the products over $CoPc^{5-}5Na^+$, for example. Fig. 4.34(a) indicates the change with

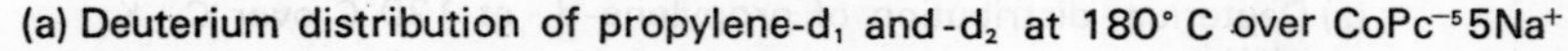

(a) Deuterium distribution of propylene-d₁ and-d₂ at 180° C over CoPc⁻⁵5Na⁺

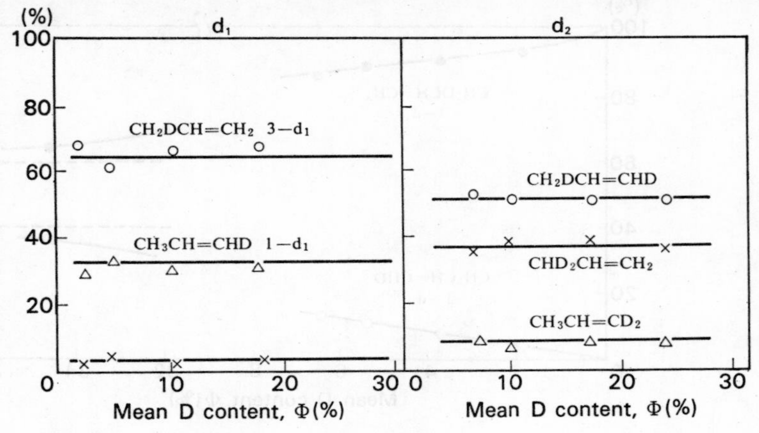

(b) Associative *iso*-propyl intermediate over $CoPc^{5-}5Na^+$

$CoPc^{5-}5Na^+$ $(d^8+\pi^4)$ $\xrightarrow{+D}$ $CH_3\text{–}CH(Co)\text{–}CH_2D$ $\xrightarrow{-H}$ $CH_3CH{=}CHD$ 40%, $CH_2DCH{=}CH_2$ 60%

$\xrightarrow{+D,\ -H}$ $CH_2DCH{=}CHD$, $CHD_2CH{=}CH_2$, $CH_3CH{=}CD_2$

Fig. 4.34 The D_2-propylene exchange reaction over $CoPc^{5-}5Na^+$, and proposed mechanism for deuterium incorporation through an *iso*-propyl intermediate.

time of the ratios of d_1 and d_2 species through the reaction over the Co-phthalocyanine^{5-} EDA complex film. A proportion of 65% 3-d_1 and 33% 1-d_1 was found in the propylene-d_1 species. The deuterium distributions of d_1 and d_2 species remain constant during the course of the reaction. These results are expected from an exchange process occurring mainly through an *iso*-propyl intermediate. If hydrogen exchange takes place only through an *iso*-propyl intermediate, the ratio of propylene-d_1 formed should be propylene-3-d_1 60% and 1-d_1 40%. The deuterium distribution of propylene-d_2 (1,3-d_2 30% and 1,1-d_2 10%) can be understood in terms of an exchange mechanism via this *iso*-propyl intermediate. Other intermediate mechanisms such as σ-allyl, π-allyl or a concerted mechanism can be excluded. If hydrogen exchange occurs through, for example, a σ-allyl intermediate, we should find 100% of 3-d_1 in the d_1 species and 100% of 2,3-d_2 in the d_2 species, regardless of reaction time. Thus, the deuterium-propylene exchange proceeds through an *iso*-propyl intermediate only over $CoPc^{5-}$-$5Na^+$ film (Fig. 4.34(b)).

Fig. 4.35 presents equivalent results for the deuterium distribution of propylene-d_1 and -d_2 in propylene-deuterium exchange over a $NiPc^{4-}4Na^+$ film. The deuterium distribution of the d_1 and d_2 species over this complex film also remained constant during the course of the reaction, but a different distribution of d_1 and d_2 species was observed over $NiPc^{4-}4Na^+$ in

(a) Deuterium distribution of propylene-d_1 and -d_2 at 180° C over $NiPc^{4-}4Na^+$

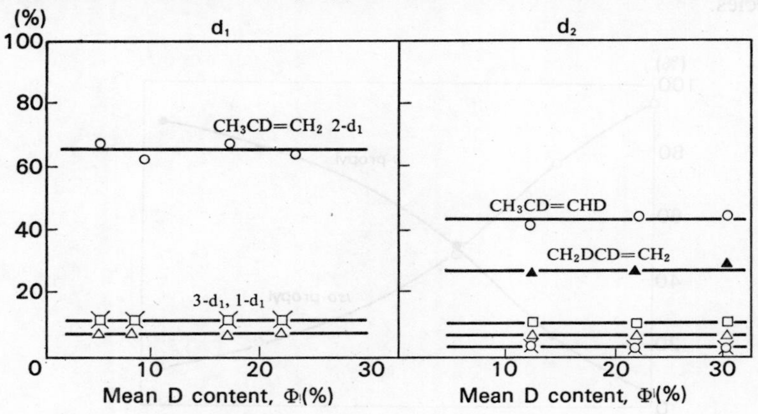

(b) Associative *n*-propyl intermediate over $NiPc^{4-}4Na^+$

$CH_3CH{=}CH_2$ + D → CH_3–CHD–CH_2 (on Ni of NiPc) → −H → 2-d_1 $CH_3CD{=}CH_2$ 100% (63%)

$NiPc^{4}4Na^+$ ($d^8+\pi^4$)

(c) Mixed intermediates in the D_2-propylene exchange over $NiPc^{4-}4Na^+$ film

n-propyl : *iso*-propyl ≒ 3 : 2

Fig. 4.35 The D_2-propylene exchange reaction over $NiP^{-4-}4Na^+$, and proposed mechanism for deuterium incorporation through mixed intermediates of *n*- and *iso*-propyl species.

comparison with that over $CoPc^{5-}5Na^+$. For the d_1 species, the results were 2-d_1 63%, 3-d_1 13% and 1-d_1 24%. This deuterium distribution is almost equal to that expected from an associative mechanism through an *n*-propyl intermediate. If the hydrogen exchange occurs through an *n*-propyl intermediate only, the propylene-d_1 formed should be 100% 2-d_1. The formation of 3-d_1 (13%) and 1-d_1 would result from an additive exchange mechanism through *iso*-propyl. We estimate from the ratio of propylene-2-d_1 in the d_1 species that the deuterium-propylene exchange proceeds through both *n*- and *iso*-propyl intermediates (3:2). The deuterium distribution of propylene-d_2 can be explained by assuming that the exchange of the second hydrogen atom of propylene-d_1 follows the same mechanism through the same mixed intermediate of *n*- and *iso*-propyl species.

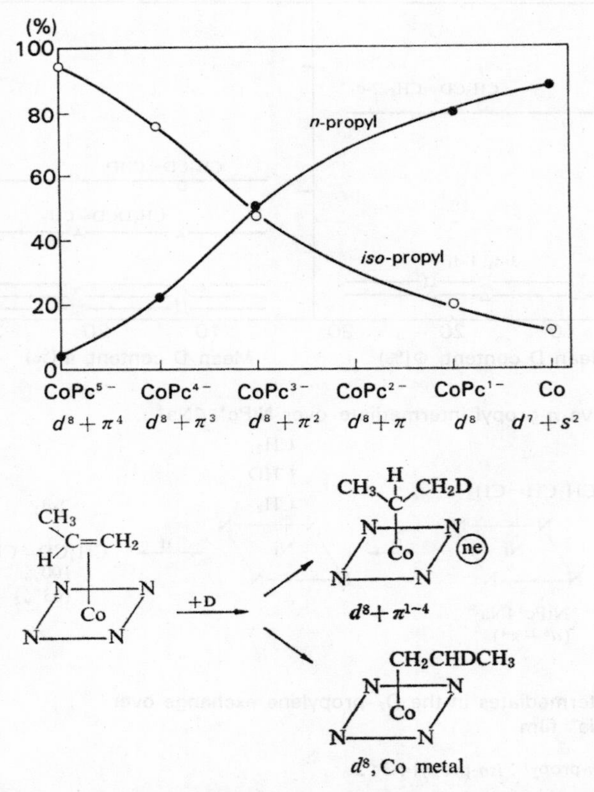

Fig. 4.36 Changes in reaction intermediates for the D_2-propylene exchange reaction over $CoPc^{n-}nNa^+$ (n=1~5) films, and electron configurations of CoPc polyanion species.

By similar procedures the adsorbed states of propylene were determined over various Co-phthalocyanine EDA complexes with sodium by changing the degree of reduction. As shown in Fig. 4.36, the d_1 species formed over each complex film varied from 3-d_1 to 2-d_1 as the main product with decreasing degree of reduction of Co-phthalocyanine, from $CoPc^{5-}$-$5Na^+$ ($d^8+\pi^4$) to $CoPc^-Na^+$ (d^8). From the ratios of propylene-2-d_1 and 3-d_1, the change was presumably from all *iso*-propyl over $CoPc^{5-}5Na^+$ to all *n*-propyl over $CoPc^-Na^+$. It has been demonstrated previously by microwave analysis that deuterium-propylene exchange proceeds through all *n*-propyl intermediate over a cobalt metal surface, having a d^8-electron structure. The ratio of *n*-propyl decreases and the ratio of *iso*-propyl increases considerably with increase of electron donation to the π-conjugate system of the Co-phthalocyanine. Thus, it was suggested that the adsorbed state of propylene on the central cobalt atom changes from *n*-propyl to *iso*-propyl due to electron donation from the polynegatively charged porphine ligand.

Naito, Ichikawa and Tamaru[59] have also observed different intermediate species for the deuterium-propylene exchange reaction over different metal-phthalocyanine EDA complexes. The ratios of the d_1 and d_2 species formed varied on changing the central metal atom (Fe, Co or Ni) in the phthalocyanine EDA complexes with sodium. Interestingly, the *iso*-propyl intermediate is observed mainly over $FePc^{4-}4Na^+$ and $CoPc^{5-}$-$5Na^+$, but *n*-propyl is the main intermediate over $NiPc^{4-}4Na^+$. All the complexes have the same electronic configuration, $d^8+\pi^4$. Table 4.11 shows that the adsorbed state of propylene over these transition metal-phthalocyanine EDA complexes is considerably different from that over

TABLE 4.11 Adsorbed states of propylene in deuterium-propylene exchange over metal-phthalocyanine EDA complexes, and electron-configuration of metal-phthalocyanine polyanions

MePc	Metal	$MePc^{n-}nNa^+(n=1\sim5)$ d^8 .... $d^8+\pi^4$
FePc	$n \gg iso$	$iso \gg n$
CoPc	$n \gg iso$	$iso \gg n$
NiPc	$n \approx iso$	$n > iso$

n-propyl: CH_3–CHD–CH_2–Me (N–N, *ne*) ← +D — CH_3–C(H)=CH_2 / Me (N–N, *ne*) — +D → *iso*-propyl: CH_3–C(H)–CH_2D / Me (N–N, *ne*)

the corresponding metal surfaces having similar *d*-electron structures. We suggest that the adsorbed state of olefins is controlled not only by the *d* character of the central metal atoms but also by electron donation from the negatively-charged phthalocyanine ligand.

4.3.2 Isomerization of butenes by EDA complexes

Naito, Ichikawa and Tamaru[59] have recently suggested that the double-bond isomerization and deuterium exchange reaction of 1-butene proceed simultaneously over phthalocyanine EDA complexes through a half-hydrogenated intermediate such as an *iso*-butenyl species. They studied the relative rates of the two reactions over $NiPc^{4-}4Na^{+}$ and $CoPc^{5-}5Na^{+}$ films and found that the rates of incorporation of deuterium in butene isomers through an *iso*-butenyl species are almost equal to those of the double-bond migration of 1-butene. They determined the rates and distributions of deuterium incorporation through the course of the isomerization by means of microwave spectroscopic analysis. The results implied that butene isomerization takes place over EDA complexes which are active for hydrogen dissociation via the *iso*-type (not *n*-type) of the half-hydrogenated butene intermediate, as follows:

$CH_3CH_2CH(CH_2D)$–MePc (ne) ←(+D)— $CH_3CH_2(H)C{=}CH_2$ π-adsorbed on $MePc^{n-}nNa^{+}$ (ne) —(+D)→ $CH_3CH_2CHDCH_2$–MePc (ne)

Double-bond migration (isomerization): −H / +H → $CH_3CH{=}CHCH_2D$ (2-butene-1-d$_1$); −H → $CH_3CH_2CH{=}CHD$ (1-butene-1-d$_1$); −H → $CH_3CH_2CD{=}CH_2$ (1-butene-2-d$_1$)

In addition, isomerizations of butenes taking place through other dissociative mechanisms have been reported over certain EDA complexes. A very high activity for the H_2–D_2 exchange reaction is seen in EDA complexes of anthracene with sodium, but the isomerization of butene is slow, suggesting that adsorption of the reacting gas influences the activity of the EDA complexes to some extent. Pines and his co-workers have reported[60] that anthracene and sodium complex is catalytically active for double-bond migration at high temperatures under high pressures. They postulated that

TABLE 4.12 Specific isomerization of 1-butene to *cis*-2-butene over various 1:2 aromatic hydrocarbon–alkali metal EDA complexes

Complex (3×10^{-3} mole)	T (°C)	Conv. % (20 hr)	*cis*-2-butene / *trans*-2-butene
$]^{2-}2Na^{+}$	160	4	> 5.6
$]^{2-}2Rb^{+}$	150	12	> 7.1
$]^{2-}2Cs^{+}$	110	6	> 9.5
	160	20	> 6
$]^{2-}2Rb^{+}$	160	5	3.5
$]^{2-}2Cs^{+}$	105	20	> 8.6
$]^{2-}2Li^{+}$	160	40	3.5

$$CH_3CH_2CH{=}CH_2 \underset{k_c}{\overset{k_t}{\rightleftarrows}} \begin{cases} CH_3CH{=}CHCH_3 \ (trans) \\ CH_3CH{=}CHCH_3 \ (cis) \end{cases}$$

the anthracene carbanion attaches to an allylic hydrocarbon, and that the transition state of a cyclic structure of butenyl anions and sodium cation participates in the isomerization.

By using stoichiometric 1:1 and 1:2 EDA complex films of polycondensed hydrocarbons with alkali metals such as potassium, it has been observed by Ichikawa and Tamaru[61] that the isomerization of 1-butene takes place slowly but selectively to *cis*-2-butene (*cis*-2-butene/*trans*-2-butene$=5\sim10$), and the isomerization rates do not depend much upon the structures of the aromatic hydrocarbons or the type of alkali metal, as suggested by the data in Table 4.12. In addition, propylene is adsorbed as a π-allyl intermediate over these EDA complexes, as shown by microwave spectrometry, and it thus appears that the isomerization of butene proceeds via intermolecular hydrogen transfer in the π-allyl intermediate, as is the case with typical solid base catalysts such as ZnO[62]

The proposed mechanism of isomerization of olefins is illustrated in Fig. 4.37, including that of the hydrogenation of olefins over EDA complexes.

Tsuda, Inokuchi *et al.*[63] have reported that double-bond migration of butene proceeds on violanthrene A–I_2 EDA complex films. As shown in Fig. 4.38, the amount of 1-butene decreased, whereas *cis*- and *trans*-2-butenes increased, during the course of the reaction at 60°C with an initial

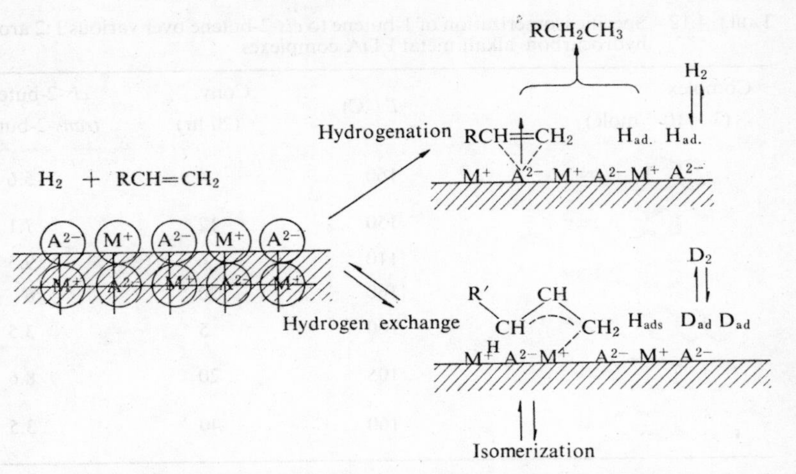

Fig. 4.37 Possible mechanisms of olefin activation over EDA complexes, for the hydrogenation, hydrogen exchange and isomerization reactions.

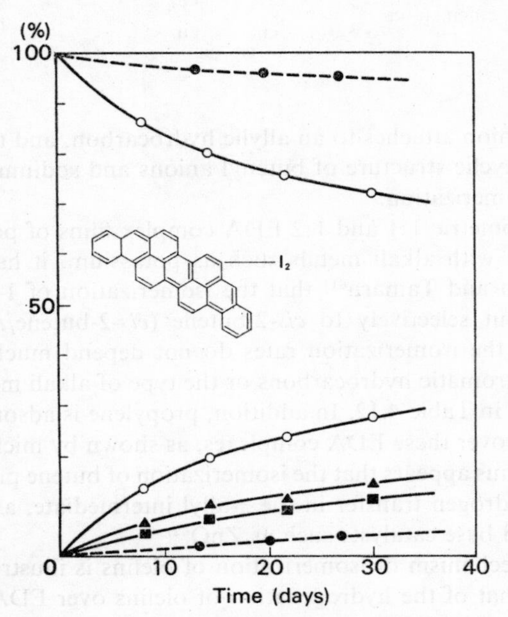

Fig. 4.38 1-Butene → 2-butene isomerization by the violanthrene A–I$_2$ charge-transfer complex. ○, Isomerization over Sample I (150 mg) at 60°C; ▲, increase of *trans*-2-butene, and ■, increase of *cis*-2-butene, over Sample I. ●, 1-Butene → 2-butene over Sample II (150 mg) at 60°C.
(Source: M. Tsuda, T. Kondow, H. Inokuchi and H. Suzuki, *J. Catalysis*, **11**, 82 (1968). Reproduced by kind permission of Academic Press, Inc., U.S.A.)

1-butene pressure of 100 mmHg. The 1-butene was transformed with approximately the same probability to *cis*-2-butene and *trans*-2-butene in the temperature range from 20 to 60°C (*cis*-2-butene/*trans*-2-butene = 1). Butene isomerization occurred only at a negligible rate with violanthrene A or iodine alone under similar reaction conditions. It was also found that the iodine content of the complex strongly affected the rate of isomerization, as shown in Fig. 4.44, where the samples I and II have the compositions (V-A)–$I_{3.18}$ and (V-A)–$I_{2.10}$, respectively. The H_2–D_2 exchange reaction did not proceed over violanthrene A–I_2 complex films at 60°C.

4.4 Ammonia Synthesis by EDA Complexes

It is extremely interesting that ammonia synthesis from nitrogen and hydrogen occurs over an Fe-phthalocyanine–Na system even at room temperature, when the catalyst is finely dispersed on active carbon.

$$3H_2 + N_2 = 2NH_3$$

Ammonia synthesis from its elements, or the fixation of nitrogen from the air, is a very well-known reaction which Haber (in Germany) first developed at the beginning of this century. Commercial catalysts, consist mainly of Fe–Al_2O_3 or Fe–K_2O–Al_2O_3 systems with other additives. The synthesis is an exothermic reaction in which four gaseous molecules (three of hydrogen, one of nitrogen) form two molecules of ammonia, resulting, according to the laws of thermodynamics (Le Chatelier's principle), in a higher yield of ammonia at lower temperatures and under higher pressures, i.e. insofar as the equilibrium point is concerned. The industrial processes of ammonia synthesis using Haber-Bosch catalysts are generally carried out at high pressures of 100–1000 atm. High temperatures of 350–500°C are employed mainly to ensure a suitable rate of ammonia formation. Accordingly, it is of interest to find effective catalysts which can accelerate the rate at which the reaction proceeds toward its equilibrium point at lower temperatures. In particular, in the case of EDA complexes, substances such as phthalocyanine and alkali metals which are not active separately towards hydrogen and nitrogen do sometimes become good catalysts on being brought into contact with each other as EDA complexes.[64)]

When each (about 100 mg or 2×10^{-4} mole) of the phthalocyanines of iron, cobalt, nickel, chlorotitanium (II), molybdenum, platinum, copper, zinc and the metal-free compound were evaporated onto a sodium film on the wall of a glass vessel, deep-colored EDA complex films were formed *in vacuo* at room temperature. (The surface area of the complex films was estimated to be less than 1 m^2 by the BET method using nitrogen.) When nitrogen gas (20 cm Hg) was introduced onto the EDA complex films of Fe-, Co-, Ti- and Mo-phthalocyanines, a considerable amount of nitrogen (1.6 ml STP) was chemisorbed in the temperature range between 25 and 240°C. Hydrogen was also adsorbed to a considerable extent over these complex films.

Mixtures of N_2 and H_2 (total pressure <60 cm Hg) with various molar ratios were circulated over Fe-phthalocyanine-sodium film in the temperature range between 25 and 260°C for 20 hr. The volume of the closed circulating system was about 140 ml and the ammonia produced was collected. An appreciable amount of ammonia was obtained at temperatures above 110°C, as shown in Table 4.13. The catalytic activity of the complex film for ammonia formation did not change appreciably in more than 10 runs. No ammonia or other product was obtained when hydrogen was admitted to the Fe-phthalocyanine complex film at above 280°C in the absence of nitrogen gas. Ammonia was also formed to some extent above 170°C over Mo-, Ti- and Co-phthalocyanine EDA complex films with sodium or potassium in the H_2–N_2 system. A negligible amount of ammonia, on the other hand, was formed over Cu-, Pt-, Zn- and metal-free phthalocyanine EDA complexes with sodium when a mixture of N_2 and H_2 was introduced onto the complex films above 200°C, even though they are markedly active for hydrogen activation reactions. A considerable amount of ammonia was adsorbed and decomposed into a mixture of N_2 and H_2 over Fe-phthalocyanine EDA complex film above 120°C. The activity of the complex films decreased considerably in the presence of oxygen.

By admitting nitrogen gas onto FePc–Na complex films prepared on a KBr disk, new absorption bands have been observed in the regions of 2196 and 2036 cm^{-1}, due probably to nitrogen adsorption onto reduced Fe in the phthalocyanine EDA complex.

TABLE 4.13 Ammonia synthesis from N_2 and H_2 over FePc–Na complex film†

N_2 (cm Hg)	H_2 (cm Hg)	T (°C)	NH_4 formed in 20 hr (ml STP)
10	30	110	0.26
10	30	170	0.92
10	30	240	3.64
10	30	260	4.60
30	10	240	1.80

† The complex film was heated at 200°C for 20 hr prior to use and a mixture was circulated with a constant rate of 12.2 ml/min.

4.5 Reduction of CO and CO_2 by EDA Complexes (Fischer-Tropsch Reactions)

Ichikawa, Sudo, Soma, Onishi and Tamaru[65] have studied the catalytic behavior of phthalocyanine and graphite EDA complexes with alkali metals in the Fischer-Tropsch reaction. Carbon monoxide (20 cm Hg) was adsorbed to a considerable extent on each EDA complex film in the temperature range between 25 and 200°C, but desorption proceeded very slowly even at higher temperatures such as 200°C. When hydrogen (20 cm Hg) was admitted at 170°C onto complex films with preadsorbed CO, a considerable amount of hydrogen was taken up and a mixture of hydrocarbons such as ethane, ethylene and propane was detected by gas chromatography in a closed, circulating reaction system (140 ml).

A mixture of CO and H_2 of various relative molar ratios reacted over each phthalocyanine EDA complex film with excess sodium in the temperature range between 25 and 260°C. Various kinds of hydrocarbons (C_1–C_5) were formed at temperatures above 90°C in 20 hr. The amount of hydrocarbons and distribution of products (in the steady state) are shown in Table 4.14. It was interesting to note that small amounts of methanol

TABLE 4.14 Preparation of hydrocarbons (C_1–C_5) over various phthalocyanine–Na EDA complexes and the graphite–Na complex

Acceptor (g)	Donor (g)	H_2 (cm)	CO (cm)	Temp. (°C)	Total amount of hydrocarbons (ml)	%†				
						C_1	C_2	C_3	C_4	C_5
Fe-Pc	Na	45	15	240	18	4	89	5	2	0
(0.2)	(0.5)	45	15	170	1.9	38	59	3	1	0
Co-Pc	Na	45	15	240	11	6	91	3	1	0
(0.2)	(0.5)	45	15	180	1.8	58	38	4	0	0
Pt-Pc	Na	46	12	220	10.2	2	82	9	6	1
(0.2)	(0.5)									
H_2-Pc	Na	45	15	240	2.4	10	84	6	0	0
(0.2)	(0.5)									
Graphite	Na	40	10	300	1.4	2	74	22	2	0
(5)	(2)									

† C_1, methane; C_2, ethylene + ethane; C_3, propylene + propane; C_4, 1-butene + *cis*-2-butene + *trans*-2-butene + butane; C_5, pentane, 1-pentene.

and ethanol, and a trace of CO_2, were detected in the products at lower temperatures such as 120°C under CO-rich conditions over Fe-, Mo-, and Pt-phthalocyanine complex films.

The reactivity of the complexes decreased appreciably by repeating the experiments, because the complexes were steadily destroyed by the water produced. The amounts of higher hydrocarbons (C_3–C_5) increased considerably at 240°C over the Fe-Pc complex film when the content of CO was increased. It was found that the amount of methane increased at lower temperatures under hydrogen-rich conditions. Similar hydrocarbons were also formed when a mixture of CO and H_2 was admitted to the graphite–Na (or K) complex in the temperature range between 200 and 350°C.

It has been suggested by Daumas and Herold[66] that graphite-potassium lamellar complexes such as C_8K and $C_{24}K$ irreversibly absorb two moles of carbon monoxide at about 100°C to yield an acetylene diol-type salt such as $K^+O^-{-}C{\equiv}O^-K^+$ among the graphite layers.

$$C_8K + 2CO \rightarrow C_8K(CO)_2$$

Carbon dioxide is similarly reduced over phthalocyanine and graphite EDA complexes with alkali metals to form mixtures of C_1–C_5 hydrocarbons with ethane as the main products. It is also interesting to note that dimethyl ether can be formed directly over a graphite-$PdCl_2$–Na system and PdPc–Na complex at 200–300°C from a mixture of CO_2 and H_2.[67]

4.6 Graphite-Alkali Metal Interlayer EDA Complexes as Hydrogen Reservoirs and Molecular Sieves

In graphite lamellar compounds, electrons are transferred from potassium (or cesium) to the graphite to form a two-dimensional polynegative giant anion and potassium cations. The compounds are gold in color (C_8K), blue ($C_{24}K$), etc. The structures of the lamellar compounds have been determined by the X-ray diffraction method. The potassium atoms are situated between the net planes, as shown in Fig. 4.39.

Active charcoal generally has a large surface area and is well-known for its marked capacity to sorb various gases and vapors. Graphite has no such ability, having a small surface area. It was recently discovered by

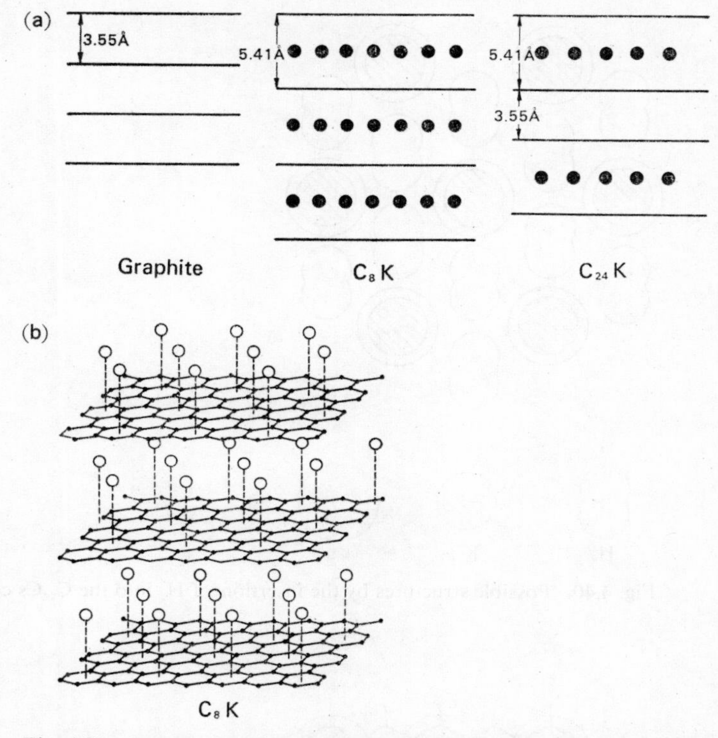

Fig. 4.39 Structures of lamellar compounds of graphite, and the interlayer distances for graphite, C_8K and $C_{24}K$ along the *c* axis.

Watanabe, Kondow, Onishi and Tamaru,[68] however, that when the graphite–potassium complex, $C_{24}K$, is exposed to hydrogen at liquid nitrogen temperature (−196°C), it absorbs several hundred times its own volume of hydrogen (hydrogen volume calculated at 0°C and 1 atm) to give the approximate composition $C_{24}KH_4$. On the other hand, very little or no nitrogen absorption is observed at this temperature. It should also be noted here that neither hydrogen nor nitrogen sorption takes place over C_8K or C_8Cs compounds, whereas $C_{24}Cs$ adsorbs both hydrogen and nitrogen (see Table 4.15; *cf.* Fig. 4.40). The hydrogen uptake is reversible and the absorbed hydrogen can easily be removed by pumping the system out. This hydrogen absorption is accompanied by a heat evolution of 2.2 kcal per mole of hydrogen taken up. When an equimolar mixture of H_2 and D_2 is contacted with $C_{24}K$ at liquid nitrogen temperature, no formation of HD is detected, showing that hydrogen is absorbed in a molecular form and does not dissociate into its atoms at such a low temperature. Moreover, it was found that H_2 is sorbed (at low pressures) about 7 times more than D_2

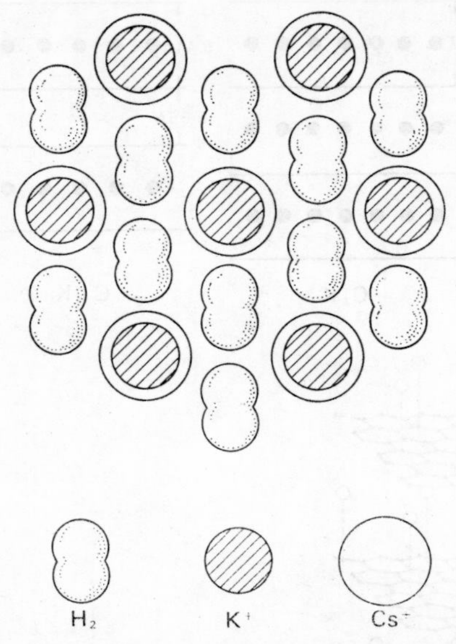

Fig. 4.40 Possible structures by the insertion of H_2 into the $C_{24}Cs$ complex.

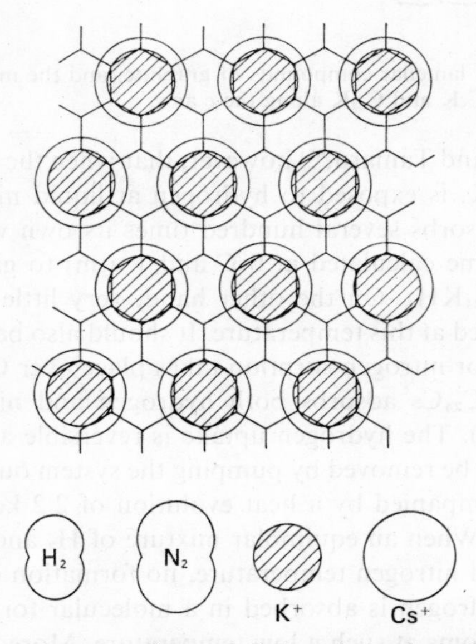

Fig. 4.41 C_8K and C_8Cs and the sizes of the H_2 and N_2 molecules and K^+ and Cs^+ ions (drawing along the *ab* planes of the complexes).

TABLE 4.15 Sorption on alkali graphites: amount of sorption at saturation (G/M)†1

	He	H_2, D_2	Ne	N_2	Ar	CH_4
$C_{24}K$	n. s.†2	2.10	n. s.	(0.7)†3	n. s.	n. s.
$C_{24}Rb$	n. s.	2.05	n. s.	1.0	1.2	0.9
$C_{24}Cs$	n. s.	2.05	n. s.	1.3	1.4	1.2

†1 G/M = Ratio of sorbed gas molecules to the number of intercalated alkali atoms.
†2 n. s. = Non-sorptive system.
†3 Attained by gradually cooling the sample in the presence of N_2.

into $C_{24}K$, suggesting that it may be possible to use the $C_{24}K$ complex as an H_2–D_2 separator in a simple procedure.

In general, nitrogen is taken up much more easily than hydrogen by active charcoal or other porous materials at low temperatures, and, consequently, the behavior of hydrogen over these lamellar compounds is characteristic. The fact that only hydrogen, and not nitrogen, is taken up in large amounts by $C_{24}K$ suggests that the $C_{24}K$ has an extremely large accessible surface area for hydrogen, but not for nitrogen. Neither C_8K nor C_8Cs has such available surface for either of these gases. In a similar sense, $C_{24}Cs$ seems to have a large accessible surface area for both hydrogen and nitrogen.

The radii of the potassium and cesium ions are approximately 1.33 and 1.69 Å, respectively. Consequently, in the structures of C_8K and C_8Cs, no free space remains for hydrogen to penetrate between the net planes, as shown in Fig. 4.41. In the case of $C_{24}K$ or $C_{24}Cs$, the number of alkali metal cations between a set of net planes becomes two-thirds that of C_8K or C_8Cs. Thus, the free space between the net planes and the alkali metal cations is increased and hydrogen molecules can be accomodated. Hydrogen molecules are thus able to penetrate into $C_{24}K$, and as they do so, the distance between the adjacent net planes increases. This reminds one rather of a sponge absorbing water and swelling.

It is also apparent from Table 4.15 that nitrogen may not penetrate between the net planes of $C_{24}K$ at liquid nitrogen temperature. As the nitrogen molecule ($r_{N_2}=1.5$ Å) is larger than K^+ ($r_{K^+}=1.33$ Å), the distance between the graphite net planes is too small for nitrogen to enter. In the case of lamellar compounds of the graphite–cesium system, on the other hand, because the Cs^+ ion ($r_{Cs^+}=1.69$ Å) is larger than the K^+ ion, the distance between the net planes becomes correspondingly greater, and the free space between the planes is sufficient for both hydrogen and nitrogen to penetrate.

It is thus clear that the size of the metal ions inserted between the net planes determines the size of the free space in the lamellar compounds, and

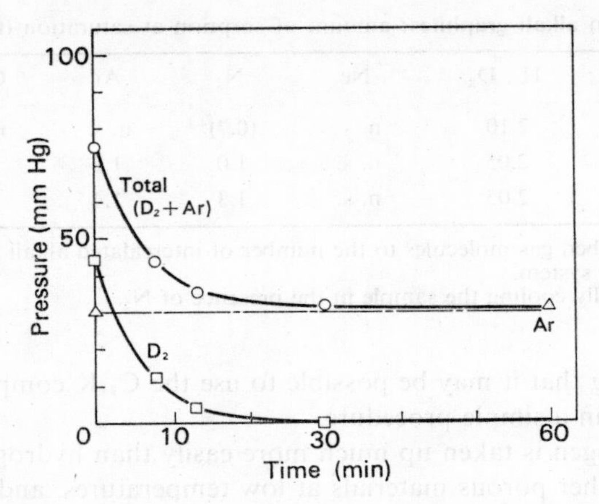

Fig. 4.42 Selective absorption of D_2 in $C_{24}K$ from a mixture of D_2 and Ar (77°K). □, D_2; △, Ar.

that various condensable gases whose size is smaller than the free space may be sorbed. The size of the free space in such lamellar compounds may be controlled by employing various alkali metals. When such a series of compounds is used, various gases of different sizes may be separated by selective sorption dependent upon their size. $C_{24}K$ does not absorb argon or methane, but $C_{24}Cs$ or $C_{24}Rb$ does to a considerable extent. Thus a separation of hydrogen and argon, for example, is possible by adsorbing the former from the mixture into $C_{24}K$. In this sense, we may regard these compounds as a new type of "molecular sieve".

Finally, it is interesting to note that when a mixture of D_2 and Ar gas was admitted onto the $C_{24}K$ complex, the D_2 was selectively absorbed by the lamellar complex, but the Ar gas remained in the gaseous phase, as shown in Fig. 4.42. In contrast to H_2, the argon atom is larger than the potassium ion, and so the distances between the graphite net planes are too small for argon to penetrate into the $C_{24}K$ complex.[69]

REFERENCES

1. R. S. Mulliken, *J. Phys. Chem.*, **56**, 801 (1952).
2. S. P. McGlymn, *Chem. Rev.*, **58**, 1113 (1958).
3. Y. Ben, Tarri Hi and J. H. Lunsford, Int. Congr. Catalysis, Preprint #109 (1972); C. Naccache and M. Che, *ibid.*, #101 (1972).
4. J. Kanthold and K. Hauffe, *Ber. Bunsenges. Phys. Chem.*, **69**, 168 (1965).
5. T. Misra, B. Rosenberg and R. Switzer, *J. Chem. Phys.*, **48**, 2096 (1968).

6. T. Seiyama and S. Kagawa, *Anal. Chem.*, **38**, 1069 (1966).
7. T. C. Waddington and W. G. Schneider, *Can. J. Chem.*, **494**, 789 (1958).
8. D. R. Kearns and M. Calvin, *J. Chem. Phys.*, **29**, 950 (1958).
9. D. R. Kearns, G. Tollin and M. Calvin, *ibid.*, **32**, 1020 (1960); D. R. Kearns and M. Calvin, *J. Am. Chem. Soc.*, **83**, 2100 (1961).
10. H. Kuroda and E. A. Flood, *J. Chem. Phys.*, **33**, 952 (1960); *Can. J. Chem.*, **39**, 1475 (1961).
11. A. N. Webb, *II Int. Congr. Catalysis*, Preprint #62 (1960).
12. S. Ross and J. P. Oliver, *J. Phys. Chem.*, **65**, 1664 (1961).
13. J. J. Rooney and R. C. Pink, *Proc. Chem. Soc.*, **142**, 70 (1961).
14. R. D. Gardner, R. L. Brandon and N. G. Nix, *Chem. Ind.*, 1363 (1958).
15. Y. Morooka and A. Ozaki, *J. Am. Chem. Soc.*, **89**, 5124 (1967).
16. R. J. Cvetanović, F. J. Duncan and W. E. Falconer, *Can. J. Chem.*, **42**, 2410 (1964).
17. E. Kh. Enikeev, L. Ya. Margolis and S. Z. Roginskii, *Dokl. Akad. Nauk, USSR*, **124**, 606 (1959).
18. A. Nakamura and S. Otsuka, *J. Am. Chem. Soc.*, **95**, 5091 (1973).
19. E. Wigner, *Z. Phys. Chem.*, ***B*23**, 28 (1933).
20. M. Calvin, E. G. Cockbain and M. Polanyi, *Trans. Faraday Soc.*, **32**, 1436 (1936); D. D. Eley and M. Polanyi, *ibid.*, **32**, 1443 (1936).
21. D. D. Eley, *ibid.*, **36**, 500 (1940).
22. L. G. Harrison and E. A. McDowell, *Proc. Roy. Soc.*, ***A*220**, 77 (1953).
23. D. D. Eley and H. Inokuchi, *Z. Elektrochem.*, **63**, 29 (1959).
24. G. J. K. Acres and D. D. Eley, *Trans. Faraday Soc.*, **60**, 1157 (1964).
25. T. Kondow, H. Inokuchi and N. Wakayama, *J. Phys. Chem.*, **43**, 3776 (1965); N. Wakayama and H. Inokuchi, *J. Catalysis*, **11**, 143 (1968); Y. Mori and H. Inokuchi, *ibid.*, **12**, 15 (1968).
26. H. Inokuchi, Y. Mori and N. Wakayama, *J. Catalysis*, **8**, 288 (1967).
27. M. Ichikawa, M. Soma, T. Onishi and K. Tamaru, *ibid.*, **6**, 336 (1966); *Bull. Chem. Soc. Japan*, **40**, 1015 (1967).
28. M. Ichikawa, M. Soma, T. Onishi and K. Tamaru, *Bull. Chem. Soc. Japan*, **40**, 1294 (1967).
29. G. Briegleb, *Angew. Chem.*, **76**, 326 (1964).
30. S. Tanaka, S. Naito, M. Ichikawa, M. Soma, T. Onishi and K. Tamaru, *Trans. Faraday Soc.*, **66**, 976 (1970); S. Tanaka, M. Ichikawa, S. Naito, M. Soma, T. Onishi and K. Tamaru, *Bull. Chem. Soc. Japan*, **41**, 1278 (1968).
31. H. Inokuchi, N. Wakayama and T. Hirooka, *J. Catalysis*, **8**, 383 (1967).
32. T. Kondow, T. Mori and H. Inokuchi, *private communication.*
33. G. Dallinga, E. L. Mackov and A. A. Verrijn Stuart, *Mol. Phys.*, **1**, 123 (1958).
34. N. H. Velthorst and G. J. Hoijtink, *J. Am. Chem. Soc.*, **87**, 4529 (1965).
35. W. G. Schneider, *NMR Chemistry Proc. Symp.* (Italy, 1964), p. 63, Pergamon.
36. M. Ichikawa, M. Soma, T. Onishi and K. Tamaru, *J. Am. Chem. Soc.*, **91**, 6505 (1969).
37. M. Ichikawa, M. Soma, T. Onishi and K. Tamaru, *Trans. Faraday Soc.*, **63**, 2012 (1967).
38. M. Ichikawa and K. Tamaru, *J. Am. Chem. Soc.*, **93**, 2079 (1971).
39. M. Ichikawa and K. Tamaru, *Z. Phys. Chem. N. F.*, **84**, 217 (1973).
40. M. Ichikawa, S. Tanaka, S. Naito, T. Nakamura, M. Soma, T. Onishi and K. Tamaru, *Bull. Chem. Soc. Japan*, **45**, 1956 (1972).
41. M. Ichikawa and K. Tamaru, *J. Chem. Soc., I. Faraday Trans.*, **69**, 1759 (1973).
42. K. Watanabe, T. Kondow, M. Soma, T. Onishi and K. Tamaru, *Nature*, **233**, 160 (1971).
43. A. R. Ubbelohde, *Proc. Roy. Soc.*, ***A*312**, 371 (1969).
44. E. Iida, M. Ichikawa and K. Tamaru, *presented for publication.*
45. M. Tsuda, H. Inokuchi and H. Suzuki, *J. Phys. Chem.*, **73**, 1595 (1969).
46. R. Taube, *Z. Chem.*, **6**, 8 (1966).

47. S. Naito, M. Ichikawa and K. Tamaru, *J. Chem. Soc. I. Faraday Trans.*, **68**, 1451 (1972).
48. M. Ichikawa, M. Soma, T. Onishi and K. Tamaru, *Bull. Chem. Soc. Japan*, **41**, 1739 (1968).
49. D. A. Dowden, N. MacKenzie and B. M. Trapnell, *Proc. Roy. Soc.*, **A237**, 235 (1956).
50. M. Ichikawa, M. Soma, T. Onishi and K. Tamaru, *Trans. Faraday Soc.*, **63**, 997 (1967).
51. K. Tamaru, *Advan. Catalysis*, **20**, 327 (1969).
52. M. Ichikawa, M. Soma, T. Onishi and K. Tamaru, partly reported in ref. 51, unpublished work.
53. H. Inokuchi, M. Tsuda and T. Kondow, *J. Catalysis*, **8**, 91 (1967).
54. M. Ichikawa, M. Soma, T. Onishi and K. Tamaru, *ibid.*, **9**, 418 (1967).
55. M. Ichikawa, M. Soma, T. Onishi and K. Tamaru, *Trans. Faraday Soc.*, **66**, 981 (1970).
56. M. Ichikawa and K. Tamaru, *J. Chem. Soc. I. Faraday Trans.*, in press.
57. T. Kondo, S. Saito and K. Tamaru, *J. Am. Chem. Soc.*, presented for publication.
58. T. Kondo, M. Ichikawa, S. Saito and K. Tamaru, *Bull. Chem. Soc. Japan*, **45**, 1580 (1972).
59. S. Naito, M. Ichikawa and K. Tamaru, *J. Chem. Soc. I. Faraday Trans.*, **69**, 685 (1973).
60. H. Pines and W. O. Haag, *J. Org. Chem.*, **33**, 328 (1958).
61. M. Ichikawa and K. Tamaru, *J. Chem. Soc. I. Faraday Trans.*, in press.
62. J. Kokes and A. L. Dent, *Advan. Catalysis*, **22**, 51 (1972); C. C. Chang and R. J. Kokes, *J. Am. Chem. Soc.*, **92**, 7517 (1970).
63. M. Tsuda, T. Kondow, H. Inokuchi and H. Suzuki, *J. Catalysis*, **11**, 81 (1968).
64. M. Sudo, M. Ichikawa, M. Soma, T. Onishi and K. Tamaru, *J. Phys. Chem.*, **73**, 1174 (1969).
65. M. Ichikawa, M. Sudo, M. Soma, T. Onishi and K. Tamaru, *J. Am. Chem. Soc.*, **91**, 1538 (1969).
66. N. Daumas and A. Herold, *Compt. Rend., Ser. C*, **266**, 373 (1968).
67. S. Naito, O. Ogawa, M. Ichikawa and K. Tamaru, *Chem. Commun.*, **1972**, 1266.
68. K. Watanabe, T. Kondow, T. Onishi and K. Tamaru, *Chem. Lett.*, **1972**, 477.
69. K. Watanabe, T. Kondow, M. Soma, T. Onishi and K. Tamaru, *Proc. Roy. Soc.*, *A*333, 51 (1973).

CHAPTER 5

The Role of EDA Complexes in Biochemical Reactions

5.1 Charge-transfer Interactions in Biology

Since Mulliken suggested in an earlier paper (1952) that charge-transfer complexes or charge-transfer interactions may play an important role in biological systems, some possible implications have been discussed in a book by Szent-Györgyi. Many workers have proposed that intra- and/or inter-charge-transfer interactions across space may be involved in the groups of certain biological macromolecules and various cellular systems, and that in some cases, the complexes are directly involved in biochemical reactions, such as in photosynthesis, phosphorylation, and redox processes with flavin-nicotinamide chains.

Recent discussion has focussed mainly on particular groups of biomolecules, including model compounds containing what are thought to be essential features. These compounds include oxidation-reduction coenzymes, particularly the pyridinium moiety of pyridinium nucleotides, the isoalloxazine moiety of flavin nucleotides, nucleic acid bases, indoles, amino acids and proteins, carotenes, quinones and porphines. Among synthetic compounds with pronounced physiological activity, carcinogenic aromatic hydrocarbons and inorganic salts, and the phenothiazine drugs have received particular attention from the viewpoint of donor-acceptor interactions, viz. "coenzyme-apoenzyme" and "receptor-site-drug" complexed models. Simple Hückel molecular orbital calculations suggest that many of these compounds fall within the category of good donors (a low positive value of β for the highest-occupied molecular orbitals) or good acceptors (see Chapter 1). Pullman and Pullman have suggested that some biomolecules are both good donors and good acceptors: such compounds involve porphines, carotenes and retinenes.

Although the capacity of many compounds of biological importance to form donor-acceptor complexes may readily be demonstrated by spectroscopic data, if suitable partners for complex formation are chosen, such donor-acceptor pairs may not *per se* be of particular biochemical interest. There are in fact relatively few cases where both compounds are of biochemical interest. Nevertheless, the vastly more complicated nature of biochemical or biological reactions has so far always been apparent in cases where the direct involvement of EDA complexes in the mechanism of such reactions has been demonstrated. Even if a correlation between some type of biological activity, such as drug-efficiency, and a property of the complexes with some second component were to be observed, it would not of necessity imply a causal relationship between the two. Both properties could derive from some other, more complicated and fundamental features.

One other property of charge-transfer complexes which is likely to be of importance only in biosystems is their color. Sometimes, the mere production of color when two biochemical (or other) reagents are mixed has been given as sole evidence for the formation of EDA complexes between them. Obviously, such an observation in itself is insufficient to justify the conclusion. From the evidence given in Chapter 2, it is apparent that the visual observation of color formation is not even a necessary condition for CT complex formation. In polar solvents, the dissociation of the complex may be so high that no CT absorption is detectable. In some cases, optical absorption which develops slowly has been assigned to arise from CT complex formation between reactants including biomolecules. It is likely that such absorptions result from the products of normal covalent chemical reactions, as discussed in Chapter 3 in connection with the work of Kosower *et al.*, etc.

Charge-transfer forces usually hold the components in a complex together in a (perhaps) specific orientation. This would act to bring large molecules such as coenzymes and porphines together with their prosthetic groups correctly aligned for specific biochemical interactions.

Many large biomolecules are in certain cases good semiconductors. One could speculate that electron or charge-transfer through highly organized systems, such as mitochondria and chloroplasts, may occur by conduction under illumination or changes in the electrostatic fields at the intersphere of the biosystems. The role of the EDA complexes or interactions might be to control and modify the electron or charge flow, as a bridge between the biochemical components, like the base of a transistor. EDA complexes in the solid state, generally, have much higher conductivity in both the dark and light than the free components.

PHOTOSYNTHESIS

Donor-acceptor complexes are considered to have important functions in promoting energy transformations in lamellar biological systems. A generalized description of the roles which they are presumed to play in processes such as photosynthesis[1] and muscle contraction has been presented by Kearns and Calvin.[2] In solution, charge-transfer absorption by a complex does not provide for energy storage since the excited species might be quenched, and the energy dispersed, by the solvent and coexistent molecules in solution. When, however, the components are arranged in layers in the solid state, the polarization resulting from the transfer of electrons from the donor to the acceptor on photoexcitation may be more long-range in character because of the diffusion of charge in each solid layer. The oxidized donor and reduced acceptor are then relatively free to function independently as chemical agents. The electrical and magnetic effects of photoexcitation of model systems, composed of solid layers of relatively simple donors and acceptors, have been discussed above in Chapters 2 and 4.

In describing the primary quantum conversion process in photosynthesis, Calvin[3] has assumed that in plant chloroplasts, chloroplast chlorophyll is laminated on one side with an electron acceptor in a lipid phase, and on the other side with a donor such as ferrocytochrome in an aqueous phase. On photoexcitation the chlorophyll transfers an electron to the acceptor. Migration of an electron from the normal chlorophyll to the vacant orbital of a chlorophyll positive ion then occurs and finally, through this charge-migration process, the chlorophyll positive ion reverts to chlorophyll by capturing an electron from the donor with the simultaneous formation of ferricytochrome. The oxidized donor serves as the oxidant and the reduced acceptor as the reductant in later steps leading to the reduction of carbon dioxide and the formation of oxygen:

$$CO_2 + H_2O \rightarrow (CH_2O)_n + O_2 \tag{5.1}$$

Recently, the role of electron carrier in the reduction phase of photosynthesis has been assigned to the iron-containing protein, ferredoxin. A three-component complex of carotene (donor-carotene-acceptor) has also been discussed as a possible medium for donor-acceptor electron exchanges in photosynthesis. Platt[4] has proposed that carotene serves as a bridge for the transfer of electrons in photosynthesis through the formation of a charge-transfer complex involving an excited molecule donor and another acceptor, D*-carotene·A. There is as yet no evidence to indicate that this complex actually participates in energy transfer in photosynthesis.

The earliest work to indicate that a porphyrin could act as a charge donor was that of Kearns, Tollin and Calvin[5]. They showed that the addition of chloranil to the surface of films of metal-free phthalocyanine caused a very marked increase in the film's dark conductivity, accompanied by a large esr signal indicating the presence of free radicals. This esr signal decreased on illumination but the conductivity increased. This may be explained by the transfer of an electron from the phthalocyanine in the dark, giving rise to the chloranil anion, and the further transfer of an electron in the light giving the double anion. Such behavior has been observed in organic charge-transfer complexes, and similar effects on addition of oxygen to metal phthalocyanines probably have the same origin.

Kearns and Calvin[2] have also studied the electric and magnetic properties of chloranil- and iodine-phthalocyanine complexes in a laminated solid arrangement, both in the dark and in the light. These studies confirmed the occurrence of a charge-transfer interaction between the two molecules (Table 5.1).

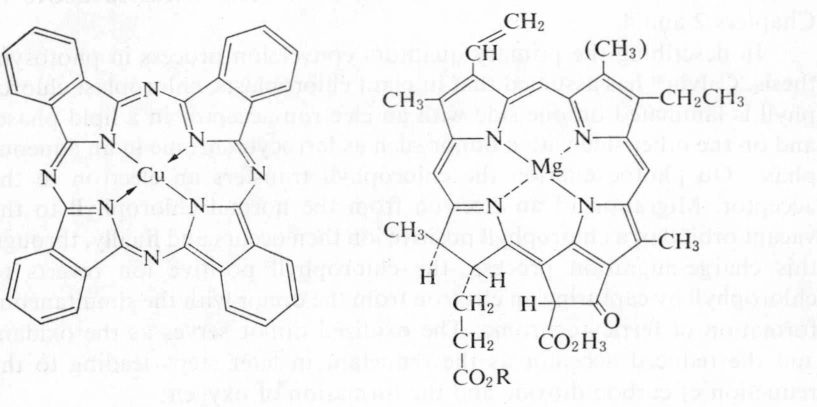

copper-phthalocyanine

chlorophyll *a* (in chlorophyll *b*, the bracketed CH_3 group is replaced by -COH)

Tollin and Green[6] have studied the effect of light on the esr signals of mixtures of chlorophyll *a* and quinones. Only in the presence of light is a large signal seen, and this arises from the semiquinone. The action spectrum for this signal is quite similar to the absorption spectrum of the chlorophyll. It would appear therefore that the excited chlorophyll is functioning as an electron donor. The authors suggest that in the case of *o*-chloranil, which has a particularly wide esr signal in the light, a complex is

TABLE 5.1 Electric and magnetic properties of metal-free phthalocyanine-acceptor lamellar systems

Donor molecule solid state ionization potential (eV)	Acceptor molecule	g-Value $\pm 0{\cdot}0002$	Line width (gauss)	No. of unpaired electrons per donor molecule	Effect of illumination on the unpaired spin concn. (approx. % change)	Photo esr decay constant, τ (sec) ($N = N_0 e^{-t/\tau}$)	Photo-current decay constant, τ (sec) ($N = N_0 e^{-t/\tau}$)
Metal-free phthalo-cyanine (4.5)	o-chloranil	2.0028	4.2S†[2]	0.002	D†[3] (10%)	65	61
	iodine	2.0030	6.7A†[2]	0.01	D†[3] (26%)	n.d.†[1]	n.d.†[1]

†[1] n.d. = no detection.
†[2] S = Symmetric line shape; A = asymmetric line shape.
†[3] D = decrease.
(Source: D. R. Kearns and M. Calvin, *J. Am. Chem. Soc.*, **83**, 2110 (1961). Reproduced by kind permission of the American Chemical Society, U.S.A.)

formed between the chlorophyll cation and semiquinone. A similar explanation is advanced for the complex with a biological quinone, coenzyme Q_6.

$$\mathrm{Ch} + h\nu = \mathrm{Ch}^*, \qquad \mathrm{Ch}^* + \text{quinone} \rightarrow (\mathrm{Ch}^+ \ldots . \text{semiquinone})^*$$
$$= \mathrm{Ch}^+ + \text{semiquinone} \rightleftharpoons \mathrm{Ch} + \text{quinone},$$

where the bracketed term represents a complex formed between the excited chlorophyll, Ch*, and the semiquinone with vibrational energy of excitation.

At low temperatures, it is suggested that a ground-state complex is formed between the chlorophyll and quinone before the formation of the excited-state complex.

$$\mathrm{Ch} + \text{quinone} \rightleftharpoons (\mathrm{Ch}\ldots\text{quinone})$$
$$(\mathrm{Ch}\ldots\text{quinone}) + h\nu = (\mathrm{Ch}^+ \ldots . \text{semiquinone})^*$$

Later work[7] on the effect of adding other acceptors to these systems has shown that the semiquinone of the most electron-negative acceptor in the system is produced on illumination. NADH caused an eightfold enhancement of the free radical concentration. This is explained by the regeneration of neutral chlorophyll from its ion by the oxidation of NADH, with the oxidized NADH then reacting with the acceptor anion to produce NADH and neutral acceptor. The action of light on mixtures of chlorophyll *a* and a biological quinone (riboflavin) is to produce the

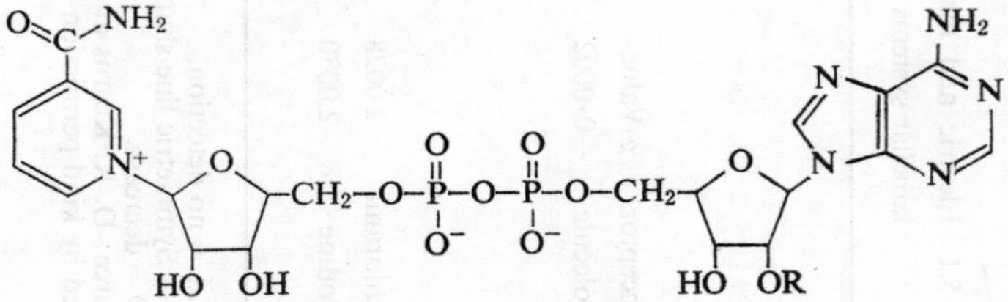

(I)

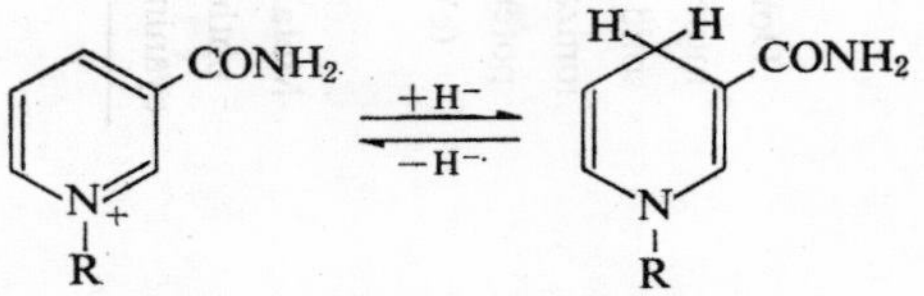

$$+\mathrm{H}^- \qquad -\mathrm{H}^-$$

(II) NAD^+ (III) NADH

riboflavin semiquinone, as detected by esr spectroscopy and ultraviolet spectrophotometry. However, there is no evidence of π-complex formation in these systems.

Boon[8] has listed effective herbicidal agents of the alkyl quaternary dipyridyl salt type as selective killers of green plants, and has proposed a good correlation between their herbicidal effectiveness and electron-accepting ability, such as the reduction potentials of the quaternary dipyridyl salts, as shown in Table 5.2. With decreasing reduction potential of

TABLE 5.2 Herbicidal efficiency and reduction potentials of quaternary dipyridyl salts

Salt	E_0 (mV)	Effective amount ($\times 10^{-5}$ M)
2,2′-dipyridyl, N^+,N^+ bridged by CH_2–CH_2, $2Br^-$	−348	1.5
HO–$(CH_2)_2$–N^+(4,4′-dipyridyl)N^+–$(CH_2)_2$–OH $2Br^-$	−408	10
CH_3–N^+(4,4′-dipyridyl)N^+–CH_3 $2Br^-$	−446	30
CH_3-substituted 2,2′-dipyridyl, –CH_3, N^+,N^+ bridged by CH_2–CH_2, $2Br^-$	−479	70
CH_3, CH_3-substituted 2,2′-dipyridyl, N^+,N^+ bridged by CH_2–CH_2, $2Br^-$	−487	100
2,2′-dipyridyl, N^+,N^+ bridged by CH_2–CH_2–CH_2, $2Br^-$	−548	500

the alkyl quaternary salts, the efficiency of killing green plants increases markedly. This suggests that electron-accepting reagents may act as components to form CT complexes with biochemically important elements in the course of, for instance, photosynthesis. In the presence of dilute solutions of herbicidal alkyl quaternary dipyridyl salts, it has been found that a considerable amount of hydrogen peroxide can be detected in green plants. This may cause inhibition of the photosynthetic cycles, probably at the photosynthesis system I in Fig. 5.1. The alkyl quaternary dipyridyl salts may trap a number of electrons in the electron-transport system from the photoexcited chlorophyll to yield bio-quinone derivatives and H_2O_2 under an oxygen atmosphere.

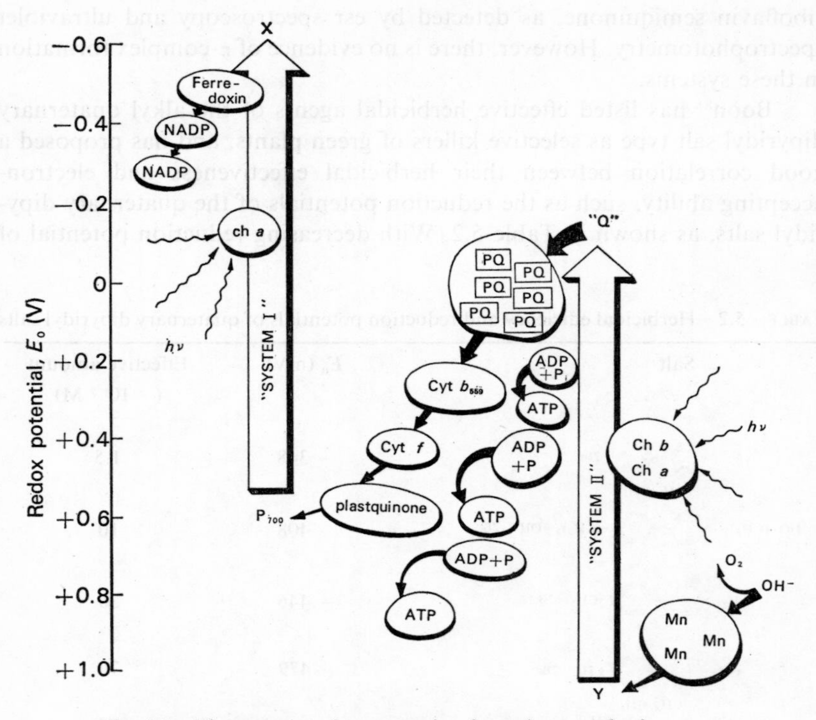

Fig. 5.1 Electron-transfer system in plant photosynthesis.

CARCINOLOGY

Among many other interesting applications of the phenomena of complex formation for interpreting the behavior of compounds of biological significance are those which are concerned with the relationship between the carcinogenic character and reactivity of aromatic hydrocarbons. The argentation constants (K_1) for reaction in equimolar water-methanol solutions, of phenanthrene, anthracene, and 22 compounds containing either the benz[*a*]anthracene (IV) or the benzo[*c*]phenanthrene (V) structure have been compared with the carcinogenic potencies of the hydro-

5 (IV) 6

(V)

carbons in mice, as determined by both skin testing and subcutaneous injection.

$$ArH + Ag^+ \rightleftharpoons ArH \cdot Ag^+; \; K_1 = [ArH \cdot Ag^+]/[ArH] + [Ag^+] \qquad (5.2)$$

The correlation is generally very good for the substituted benz[*a*]anthracenes but not for the substituted benzo[*c*]phenanthrenes. Anthracene and phenanthrene are themselves not carcinogenic but both show appreciable tendencies to undergo complexation with Ag^+. The carcinogenicities of the benz[*a*]anthracenes also correlate reasonably well with the osmium tetroxide oxidation rates of the hydrocarbons[9] and with the electronic charges, as estimated by quantum mechanical methods, at the bonds (5,6) with highest electron density.[10] The relative stabilities of 2,4,7-trinitrofluorenone complexes[11] of these benz[*a*]anthracenes are less obviously related to the carcinogenic potencies. Certainly, the interactions of this acceptor are not localized at the 5, 6 positions of the donors, as would appear to be the case for silver ion. The carcinogenic properties of the benzo[*c*]phenanthrenes are apparently dependent on some factor other than nucleophilic character. Clearly, when a sufficiently large number of hydrocarbons (both biologically active and inactive) are compared, it becomes apparent[12] that there is no simple correlation between carcinogenic activity and the electron-donor properties of the hydrocarbons (Table 5.3). In this table, the k values of the aromatic hydrocarbons may correspond to the relative donor (or acceptor) levels interacting with the biomolecules.

There is a similar lack of evidence[13] to support the hypothesis of a *direct* connection between carcinogenicity and electron affinity amongst the aromatic hydrocarbons.[14] Arguments that the electron-accepting ability should fall within a certain range have been criticized[15] for theoretical reasons; in any case, no strong correlation is observed if a sufficiently large number of aromatic hydrocarbons is taken into consideration. The suggestion that the hydrocarbon carcinogens must be simultaneously good electron donors and good electron acceptors has been made,[16] but this postulate has met with considerable criticism.[17] For example, it has been pointed out that for unsubstituted aromatic hydrocarbons there is inevitably a correlation between their electron-donating and electron-accepting properties. Again, when a large group of aromatic compounds is taken, any apparent correlation between such properties and the carcinogenicity shown by some smaller and somewhat selected group of compounds disappears. Birks[18] has suggested that the biochemical action of the hydrocarbon carcinogens may be through an energy-transfer process from an excited state of the protein moiety of a hydrocarbon-protein complex to

TABLE 5.3 Electron donor (or acceptor) properties of hydrocarbons as measured by the absolute value of the coefficient k of their highest-filled (or lowest-empty) molecular orbital†[1]

Compound†[2]	k	Carcino-genicity
Benzene	1	−
Triphenylene	0.684	−
Naphthalene	0.618	−
Phenanthrene	0.605	−
Benzo[*c*]phenanthrene (3,4-benzphenanthrene)	0.566	+
Dibenzo[*e,l*]pyrene (1,2,6,7-dibenzpyrene)	0.555	−
Dibenzo[*a,g*]phenanthrene (1,2,5,6-dibenzphenanthrene)	0.550	+
Coronene	0.539	−
Dibenzo[*c,g*]phenanthrene (3,4,5,6-dibenzphenanthrene)	0.535	−
Dibenzo[*a,c*]phenanthrene (1,2,3,4-dibenzphenanthrene)	0.532	+
Chrysene	0.520	−
Picene	0.501	−
Dibenz[*a,c*]anthracene (1,2,3,4-dibenzanthracene)	0.499	−
Benzo[*e*]pyrene (1,2-benzpyrene)	0.497	−
Dibenz[*a,j*]anthracene (1,2,7,8-dibenzanthracene)	0.492	+
Dibenz[*a,h*]anthracene (1,2,5,6-dibenzanthracene)	0.473	++
Benz[*a*]anthracene (1,2-dibenzanthracene)	0.452	±
Pyrene	0.445	−
Benzo[*a*]perylene (1,2-benzperylene)	0.439	−
Pentaphene	0.437	−
Dibenzo[*b,g*]phenanthrene (2,3,5,6-dibenzphenanthrene)	0.419	−
Anthracene	0.414	−
Dibenzo[*a,h*]phenanthrene (1,2,6,7-dibenzphenanthrene)†[3]	0.405	−
Dibenzo[*a,l*]pyrene (1,2,3,4-dibenzpyrene)	0.398	++
Dinaphtho[1,2-*b*; 1,2-*k*]chrysene (4,5,10,11-di-1′,2′-naphthochrysene)	0.383	−
Benzo[*a*]pyrene (3,4-benzpyrene)	0.371	+++
Dibenzo[*a,l*]naphthacene (1,2,9,10-dibenznaphthacene)	0.361	−
Anthra[1,2-*a*]anthracene (1′,2′-anthra-1,2-anthracene)	0.360	−
Dibenzo[*a,j*]naphthacene (1,2,7,8-dibenznaphthacene)	0.358	−
Dibenzo[*a,c*]naphthacene (1,2,3,4-dibenznaphthacene)	0.356	−
Dibenzo[*b,k*]perylene (2,3,8,9-dibenzperylene)	0.356	−
Anthra[2,1-*a*]anthracene (2′,1′-anthra-1,2-anthracene)	0.348	−
Perylene	0.347	−
Dibenzo[*a,i*]pyrene (3,4,9,10-dibenzpyrene)	0.342	+++
Benzo[*a*]naphthacene (1,2-benznaphthacene)	0.327	−
Dibenzo[*a,h*]pyrene (3,4,8,9-dibenzpyrene)	0.303	+++
Naphtho[2,3-*a*]pyrene (2′,3′-naphtho-3,4-pyrene)	0.303	−
Naphthacene	0.295	−
Anthanthrene	0.291	−
Dinaphtho[2,3-*a*; 2′,3′-*i*]pyrene (3,4,9,10-di(2′,3′-naphtho-pyrene)	0.273	−
Pentacene	0.220	−
Naphthodianthrene	0.177	−

†[1] The smaller k, the better are the donor (and acceptor) properties of the hydrocarbon.
†[2] For alternative names, see Compound Index.
†[3] Described as 2,3,7,8-dibenzphenanthrene in ref. 17.
(Source: R. Foster, *Organic Charge-Transfer Complexes*, p. 353, 1969. Reproduced by kind permission of Academic Press Inc. (London) Ltd., England.)

the hydrocarbon in an excited state. Some problems arising from this mechanism have been discussed by Pullman.[19]

The carcinogenic properties of certain aromatic hydrocarbons may in fact be directly dependent on their chemical reactivity, which is far greater than was at one time believed. For example, covalent bonding to protein *in vivo* can occur. Various hydrocarbons, when painted on mouse skin, rapidly become attached to protein by strong chemical bonds.[20] There appears to be a correlation between this chemical reactivity and the ability of the hydrocarbon to form skin tumors.[21]

Degradative experiments whereby 2-phenylphenanthrene-3,2′-dicarboxylic acid (VI) is obtained from dibenz[*a*, *h*]anthracene-protein adduct,[22] for example, show that the reactive center of this hydrocarbon is in the

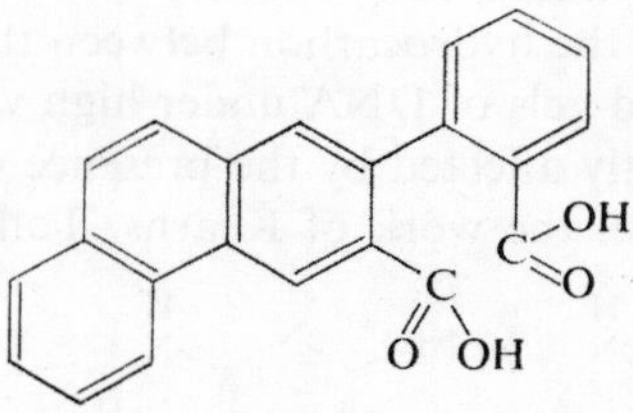

L-region

K-region

(VI) 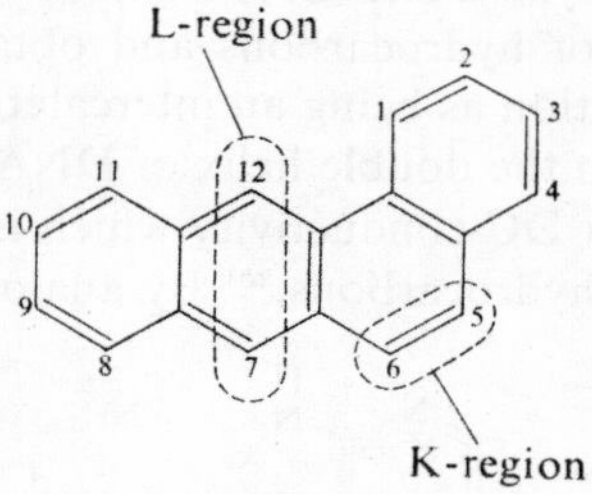 (VII)

5,6-position, the so-called K-region (see structure (VII)). This is convincing evidence in support of the proposal that the localized reactivity in this region of benz[*a*]anthracenes is important for their carcinogenicity.[23] The estimates of bond order, free valence and localization energy indicate a high reactivity in the K-region.[24] Reaction with osmium tetroxide is known to occur across this bond.[25] In order to explain cases such as unsubstituted benz[*a*]anthracene, which has a reactive K-region but is not carcinogenic, it is suggested that the reactivity in another region of the molecule, the L-region (see structure (VII)), must also be low for the compound to be biologically active. Reasonable correlations have been obtained on the basis of this assumption.

Mutagenesis

Although proteins are considered as probably the most important site for interaction with carcinogens, another possible site of interaction is at a nucleic acid. The biochemical mechanisms may have features common to the phenomenon of chemical mutagenesis, which almost certainly involves nucleic acids.

Booth and Boyland[26] have shown that DNA in the form of sodium deoxyribonucleate solubilized various polycyclic aromatic hydrocarbons in a manner similar to the bases, with the interesting difference that whereas the molar ratio of purine to hydrocarbon in the precipitated complex decreased with increasing concentration of purine, the molar ratio of DNA to the hydrocarbon increased with increasing concentration of hydrocarbon. The fluorescence maxima of the hydrocarbons were red-shifted and decreased in intensity as with the purines. The solubilizing effect of DNA on polycyclic aromatic hydrocarbons has also been reported by Liquori *et al.*[27] and T'su and Lu,[28] who showed that denaturation of the DNA, i.e. breaking down of the double-stranded helix form into a single-stranded molecule, increased the solubilizing effect.

Boyland and Green[29] have carried out similar observations on a wide range of hydrocarbons and obtained similar results. They interpret the interaction as being an intercalation of the hydrocarbon between the base pairs in the double helix of DNA. Solid gels of DNA under high vacuum show a DC conductivity which is greatly affected by the presence of aromatic hydrocarbons.[30] By analogy with the work of Kearns, Tollin and

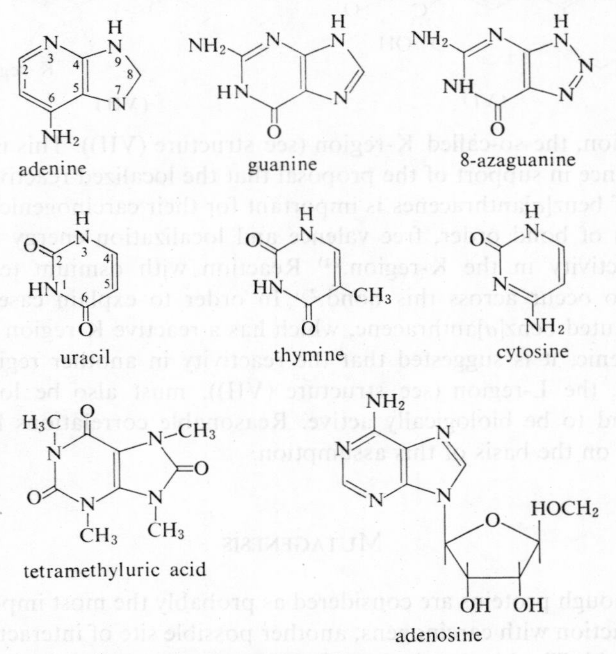

Fig. 5.2 Some purines and pyrimidines.
(Source: M. A. Slifkin, *Charge-Transfer Interactions of Biomolecules*, p. 77, 1971. Reproduced by kind permission of Academic Press Inc. (London) Ltd., England.)

Calvin,[5] in which the addition of electron acceptors to a porphyrin caused an increase in conductivity (and was directly attributed to the presence of charge-transfer states acting as shallow trapping levels for the conduction electrons, thus increasing the conductivity), it is suggested that the hydrocarbons behave as electron acceptors and DNA as an electron donor.

A large portion of the work on biological charge-transfer has been carried out with purines and pyrimidines. This can undoubtedly be attributed to the strong interest over the last few years in the role of nucleic acids in genetics. Five purines and pyrimidines, i.e. guanine, adenine, uracil, thymine and cytosine (Fig. 5.2), constitute the base pairs of DNA and RNA. However, other purines and pyrimidines do have biological importance. Caffeine, for example, is a well-known stimulant.

The earliest work in this area was carried out to determine whether purine or pyrimidine complexes can be formed with polycyclic aromatic hydrocarbons, many of which are potent carcinogens, at least to small rodents if not to human beings. As these hydrocarbons form 1:1 charge-transfer complexes with conventional organic acceptors, it was natural that workers in this field should consider the possibility of charge-transfer complexing with biological molecules.

As early as 1938, well before the idea of charge-transfer complexing was suggested, Brock and his co-workers[31] had shown that caffeine exhibits a marked solubilizing effect on various polycyclic compounds in water. Weil-Malherbe measured the solubilizing power of different purines and found that it increased with increasing *N*-methylation, so that 1,3,7,9-tetramethyluric acid was the best solubilizing agent among the purines studied (Table 5.4). Weil-Malherbe[32] was also able to isolate crystalline compounds containing purines and hydrocarbon in well-defined molecular ratios. The tetramethyluric acid–purine mixed crystal showed a 1:1 stoichiometry. (The tetramethyluric acid–(3,4-benzopyrene) mixed crystal has a 2:1 stoichiometry.) These results of Weil-Malherbe were subsequently confirmed by the X-ray crystallographic studies of Liquori *et al.*,[33] who demonstrated not only that the stoichiometry was 1:1 but also that the two molecules were arranged in the crystal in a sandwich configuration, similar to those found in crystals of well-known charge-transfer complexes. The intermolecular separation was 3.4 Å. The crystal structure of the tetramethyluric acid–pyrene complex is illustrated in Fig. 5.3.

Work on a series of hydrocarbon-purine complexes has been carried out by workers at the Chester Beatty Hospital. Booth and Boyland[26] and others[34] have examined the solubilities of various aromatic compounds in aqueous solutions. In common with earlier workers, they observed marked increases in solubility in purine solutions as compared to water. They were also able to isolate from these solutions solid CT complexes or molecular

TABLE 5.4 The solubility of polycyclic hydrocarbons in aqueous purine and pyrimidine solutions

Purine or pyrimidine	Purine or pyrimidine concn. (μM)	Hydro-carbon solubilized (μM)	M.R. purine / hydro-carbon	M.R. (caffeine)† × 100 / M.R. (purine)
Benzo(a)pyrene				
6-Dimethylaminopurine	12,250	0.374	32,800	10.1
6-Methylaminopurine	6,710	0.078	86,000	5.4
	5,000	0.047	106,400	5.0
Guanine (in N H_2SO_4)	10,000	0.115	87,000	4.5
Guanosine	3,400	0.018	189,000	3.5
Hypoxanthine	5,000	0.029	172,500	3.2
Adenine	6,000	0.021	286,000	1.8
Inosine	6,000	0.020	300,000	1.7
Adenosine	10,000	0.031	323,000	1.3
Orotic acid (pH 11·8 (NaOH))	5,000	0.007	714,000	0.75
Thymidine	30,000	0.080	375,000	0.54
Cytidine	12,070	0.013	928,000	0.34
Uracil	30,000	0.007	4,287,000	0.05
Tryptophan	50,000	<0.48	—	<1.0
Urea	6 M	0.388	15,466,000	—
Pyrene				
Guanine (in N HCl)	10,000	4.43	2,260	11.7
Adenine	6,000	0.65	9,230	3.8
Tryptophan (pH 6•5)	50,000	0.8	62,500	0.13
DPN	20,000	5.16	3,880	3.97
3-*Fluoro*-10-*methyl*-1,2-*benzanthracene* (water solubility 0.019 μM)				
Caffeine	5,000	0.19	26,300	—
Caffeine	60,000	18.33	3,270	—
4-*Fluoro*-10-*methyl*-1,2-*benzanthracene* (water solubility 0.019 μM)				
Caffeine	5,000	0.74	6,710	—
Caffeine	60,000	41.7	1,440	—

† M.R.=molar ratio.
(Source: E. Boyland and B. Green, *Brit. J. Cancer*, **16**, 347 (1962). Reproduced by kind permission of H. K. Lewis & Co. Ltd., England.)

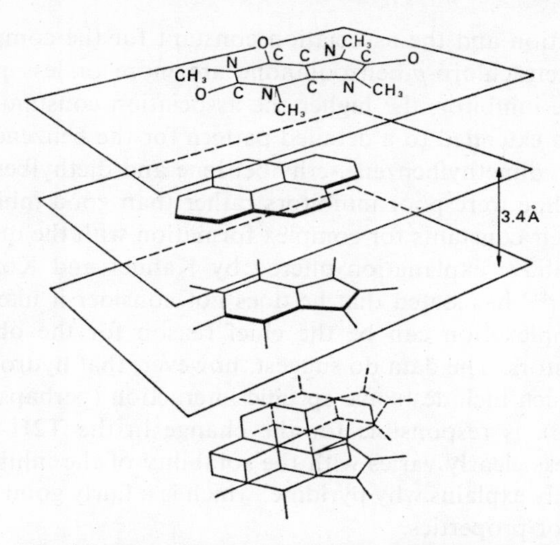

Fig. 5.3 Crystal structure of the tetramethyluric acid–pyrene CT complex: schematic drawing of the perpendicular separation of the molecules in one stack through the unit cell.
(Source: F. De Santis, A. G. Giglio, A. M. Liquori and A. Ripamonti, *Nature*, **191**, 900 (1961). Reproduced by kind permission of Macmillan Journals, Ltd., England.)

compounds containing the components in simple stoichiometry. The absorption spectra of the hydrocarbons in purine solutions are similar in form to those in 50% aqueous ethanol, but are shifted slightly to longer wave lengths and give a lower absorbance. The fluorescence of the hydrocarbons is quenched in the presence of acidified caffeine solution and in tetramethyluric acid at all pH's, as also found by Weil-Mahlherbe.

Miscellaneous Results for EDA Interactions in Biology

It is thus virtually certain that charge-transfer interaction and reactions which follow from charge-transfer complex formation are significant biochemically, although this area has not yet been explored sufficiently for any broad statement on the biochemical or biological importance of charge-transfer complex formation to be made. One of the most fascinating recent proposals for a possible intervention of charge-transfer complexation in a biological process is that of Kanner and Kozloff,[35] who suggested that the inhibition of infection of *E. coli* by T2H bacteriophage caused by indole and other compounds may result from charge-transfer complexation by the inhibitor with the sites on the phage. The evidence for this hypothesis is that the inhibition and the association constant for the complex of the inhibitor with tetrachloro-*p*-benzoquinone are more or less parallel (i.e. the stronger the inhibitor, the higher the association constant). This parallelism was even extended to a detailed pattern for the benzene series (e.g. methylbenzene, dimethylbenzene, ethylbenzene and diethylbenzene). Indazole and quinoline were poor inhibitors rather than good inhibitors, as expected from their constants for complex formation with the quinone.

Although the above explanation offered by Kanner and Kozloff is interesting, Kosower[36)] has stated that he does not consider it likely that charge-transfer complexation can be the chief reason for the observed activity of the inhibitors. The data do suggest, however, that hydrophobic bond formation, which includes some specific interaction (perhaps of the charge-transfer type), is responsible for the change in the T2H phage. Inhibitor effectiveness clearly varies with the solubility of the inhibitor in water, and this readily explains why pyridine, which is a fairly good donor, is lacking in inhibitor properties.

5.2 EDA Complexes of Biomolecules

Since 1950, charge-transfer complexes have been postulated in a variety of biochemical systems. Some of the major groups of substances involved are discussed in this section. Although the capacity of many compounds of biochemical importance to form charge-transfer complexes may be readily demonstrated when partners suitable for complex formation are selected, such donor-acceptor *pairs* may not *per se* be of particular biochemical interest. There are in fact relatively few examples where both components are of great biological interest.

Amino acids and proteins are abundant in nature and are of fundamental biological importance. However, very little work appears to have been done on these molecules in this respect, with the exception of tryptophan and tyrosine, a fact which undoubtedly arises from the generally accepted, if perhaps not fully acknowledged view, that only charge-transfer interactions involving π electrons are likely to be of biological importance. There are about 24 naturally occurring amino acids, the vast majority of which do not possess π electrons. However, the amino acids do possess an amino group ($-NH_2$) (see Fig. 5.4) which has lone-pair electrons located on the nitrogen, and these can act as σ donors to form charge-transfer complexes with various aromatic acceptors *in vitro*.

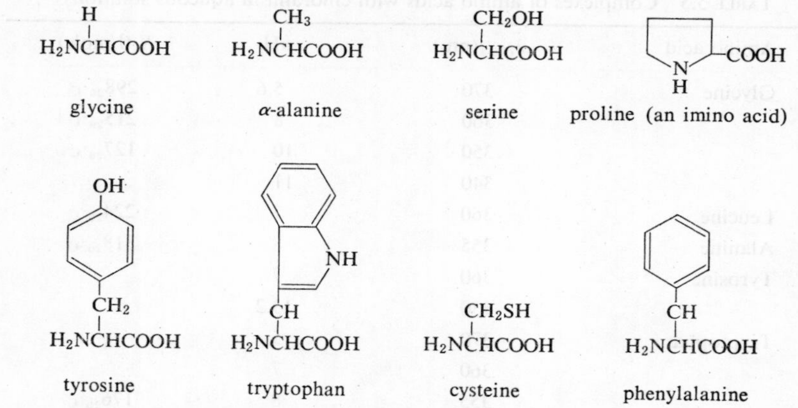

H
$H_2NCHCOOH$
glycine

CH_3
$H_2NCHCOOH$
α-alanine

CH_2OH
$H_2NCHCOOH$
serine

COOH
N
H
proline (an imino acid)

OH
CH_2
$H_2NCHCOOH$
tyrosine

NH
CH
$H_2NCHCOOH$
tryptophan

CH_2SH
$H_2NCHCOOH$
cysteine

CH
$H_2NCHCOOH$
phenylalanine

Fig. 5.4 Some common amino acids.

A series of studies has been carried out on the interaction of amino acids and proteins with the well-known organic electron acceptor chloranil (tetrachlorobenzoquinone) (see Table 5.5). This molecule was chosen as a model for the biological quinones. Spectral studies of solutions of mixtures of amino acids and chloranil in aqueous ethanol revealed some interesting changes. The spectrum of chloranil in aqueous ethanol has a strong absorbance at *ca.* 290 nm. On adding an amino acid, this peak decreases in intensity and shifts slightly, to *ca.* 295 nm in the case of phenylalanine. A new peak also appears at a longer wave length, between *ca.* 350 nm and 390 nm dependent on the pH (see Fig. 5.5).

Solid complexes of amino acids and chloranil have been prepared by evaporating down aged solutions in a rotary evaporator at low temperatures under reduced pressure. Dark brown solids are produced whose infrared spectra in KBr disks have been studied. These solids show the spectrum of the unionized amino acid, unlike that of the amino acid in the free state, which is $^{-}OOCCHRNH_3^{+}$. The spectrum for the complexed chloranil is the same as that of chloranil complexed with hydroquinone, i.e. a classical charge-transfer complex. Complexed chloranil shows a shift of the carbonyl band from 1690 to 1633 cm^{-1}. This provides unequivocal evidence that amino acids form 1:1 charge-transfer complexes of quinhydrone type with chloranil. Moreover, tryptophan and tyrosine are normally regarded as good π-electron donors, but also behave as n-donors in complexing with chloranil.

The addition of such amino acids to riboflavin in a fully oxidized form causes the absorption spectrum to change to that of a partially reduced form. Conversely, the presence of the amino acids in the riboflavin solu-

TABLE 5.5 Complexes of amino acids with chloranil in aqueous solution†

Amino acid	λ_{max}(nm)	*p*H	K_c(l/mole)
Glycine	370	5.6	$298_{20°C}$
	360	8	$215_{20°C}$
	350	10	$127_{20°C}$
	340	11	
Leucine	360	7	$224_{20°C}$
Alanine	355	8	$318_{20°C}$
Tyrosine	360	7	
	330	13.2	
Tryptophan	370	5	
	360	7	
	355	8	$176_{20°C}$
	340	10.4	
	330	12	*ca.* $0_{20°C}$
Proline	330	1	
	330	2	
	330	4	$21_{22°C}$
	375	6	
	375	7	$11_{22°C}$
	370	9	$154_{20°C}$
	(forward reaction rate 1.81×10^{-2} M^{-1} sec^{-1}, back reaction rate 1.2×10^{-4} sec^{-1})		
	340	10.2	
	355	12	
	335	13	

† Proline spectra obtained in Gallenkampf buffer; all others in Burroughs Wellcome buffer.
(Source: M. A. Slifkin, *Charge-Transfer Interactions of Biomolecules*, p. 57, 1971. Reproduced by kind permission of Academic Press Inc. (London) Ltd., England.)

tions protects the riboflavin against further reduction. Beinert[37] has proposed that the reduction of riboflavin is a two-electron process, i.e.

$$\text{oxidized riboflavin} \longleftrightarrow \text{semiquinone riboflavin} \longleftrightarrow \text{reduced riboflavin.}$$

The action of the amino acids may be to stabilize the semiquinone by one-electron donation to the oxidized flavin on forming a complex, although this opinion has been challenged by Kosower[38] who interprets the results as due to hydrophobic bonding of the amino acids to the flavin, the spec-

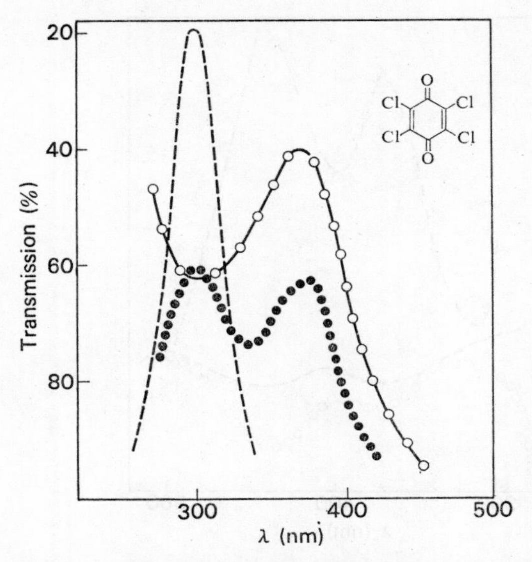

Fig. 5.5 Absorption spectra of chloranil (8×10^{-5} M) (---), chloranil (8×10^{-5} M)+glycine (10^{-2} M) (○), and chloranil (8×10^{-5} M)+phenyl-alanine (10^{-3} M) (●), all in 50% ethanol.
(Source: J. B. Birks and M. A. Slifkin, *Nature*, **197**, 42 (1963). Reproduced by kind permission of Macmillan Journals, Ltd., England.)

tral changes being simply perturbation effects. As the spectrum gives the same isobestic points as those reported by Beinert for riboflavin during the process of reduction, the first theory is regarded as the more probable, especially as the amino acids are known electron donors and the flavins good acceptors (Fig. 5.6)

A series of investigations has been carried out by Yagi and his collaborators[39)] on the interaction of the flavoprotein D-amino acid oxidase with amino acids. A mixed solution of the protein with D-lysine under anaerobic conditions yielded a purple complex with an absorption maximum at 550 nm. On adding oxygen, the purple complex changed color to green with an absorption maximum at 630 nm. Neither of these complexes gives rise to esr signals (Fig. 5.7).

Similar purple complexes have been isolated from mixtures of D-amino acid oxidase and D-proline and D-alanine. These complexes all have a 1:1 stoichiometry. The absorption maximum of the alanine complex lies at a somewhat shorter wave length than that of the proline complex, which in turn lies at a shorter wave length than that of the lysine complex. The ionization potential of the amino acid proline is about 0.27 eV smaller than that of alanine, as given by the difference between the maxima of the purple

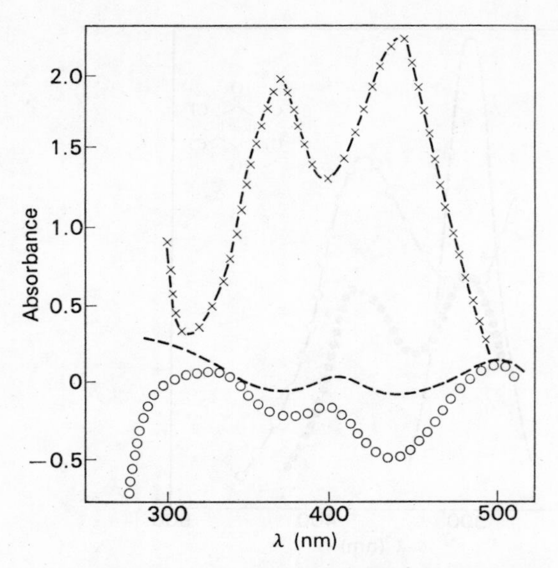

Fig. 5.6 Absorption spectrum of riboflavin (RFN) in aqueous buffer (×); and difference spectra of RFN+phenylalanine *vs.* RFN (○), and RFN+leucine *vs.* RFN (---).
(Source: M. A. Slifkin, *Nature*, **197**, 275 (1963). Reproduced by kind permision of Macmillan Journals, Ltd., England.)

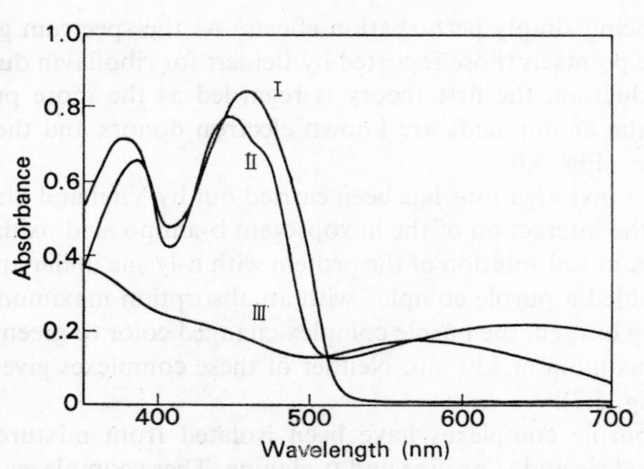

Fig. 5.7 Changes in the spectrum of D-amino acid oxidase on reaction with D-lysine. I: Oxidized state of D-amino acid, 7.44×10^{-5} M with respect to FAD. II: I was mixed with 1.5×10^{-4} M D-lysine under aerobic conditions; "green complex". III, I was mixed with 1.5×10^{-4} M D-lysine under anaerobic conditions; purple complex.
(After K. Yagi, A. Kotaki, M. Naoi and K. Okamura, *J. Biochem.*, **60**, 236 (1966).)

bands converted to energy. Slifkin and Allison[40] have shown that the difference between the ionization potentials of these compounds is 0.27 eV as determined from the contact charge spectra in the presence of oxygen.

The above purple complexes do not give rise to an esr signal. However, on standing in the dark for several days an esr signal does appear and is accompanied by a change in the absorption spectrum, the peak at 550 nm being replaced by a peak at 492 nm believed to be due to the conversion of the original purple complex to the semiquinoid enzyme. There is also a marked difference in spectrum between the purple complex and that of the semiquinone obtained by partial reduction of the enzyme.

The above results are compatible with the formation of n-π charge-transfer complexes between the lone-pair electrons on the nitrogen in the amino acid and the isoalloxazine ring of the enzyme. The suggested structure of the complexes is $FAD^{-}\cdot NH_2^{+}RCHCOO^{-}$. This is a similar structure to the amino acid–chloranil complex, which has been confirmed by infrared spectroscopy. However, in solution, the purple complexes are hydrolyzed. Ionization of the complex is largely prevented by the hydrophobic environment of the enzyme.

Yagi *et al.* have further described the purple complex as an inner complex, i.e. one which is primarily dative in the ground state, and the green complex as an outer complex, i.e. one where the ground state possess little dative character.

Isenberg and Szent-Györgyi[41] and Harbury and Foley[42] have obtained evidence for complexing between flavins and a variety of other compounds including tryptophan, serotonin, caffeine, *p*-methoxycinnamate ion, and anthraquinone-1-sulfonate ion. Although the change in the riboflavin (or

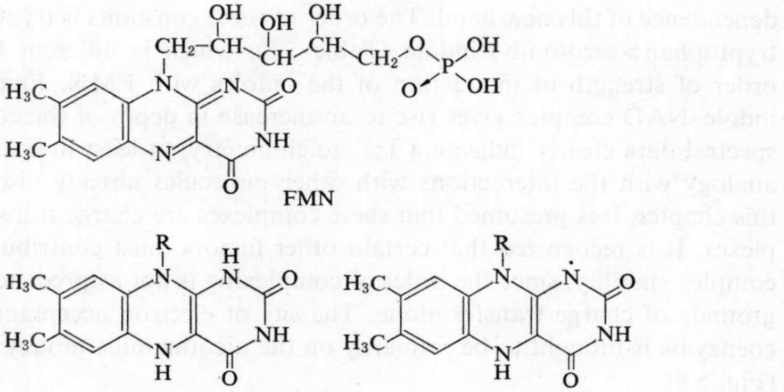

FMN

$FMNH_2$ $FMNH^{-}$

Fig. 5.8 Flavin mononucleotides: the oxidized, reduced, and semiquinone forms, respectively.

other flavin) was sufficient to enable dissociation constants to be measured, in no case was a charge-transfer band found. Radda and Calvin[43] have searched for the charge-transfer complex of NADH and FMN, reported by Isenberg, Baird and Szent-Györgyi[44] to complex and yield the radical obtained by adding an electron to FMN, the so-called flavin semiquinone ($FMN^{\overline{\cdot}}$) Fig. 5.8).

No evidence for charge-transfer complex formation was however found in either the absorption or fluorescence, although a dark reaction which gave rise to FMN was observed.

With the present paucity of quantitative information on donor-FMN complexes and their possible connection with mental activity through the role of flavin in amino acid oxidases, it would be presumptuous to do more than simply cite the suggestion by Isenberg and Szent-Györgyi[41] and by Popov and his co-workers[45] that such complexes may play a role in mental activity. The latter suggestion arose in the course of discussing research on pentamethylenetetrazole–iodine complexes. (Metrazole is used in shock therapy and as a stimulant.)

NAD and Coenzyme Models

Alivisatos *et al.*[46] have examined the nonenzymatic interaction of various indoles with NAD. The addition of indole to NAD solutions causes the appearance of a bright color due to a marked increase of absorption in the 400–500 nm region. This color shows a pH dependence, increasing markedly at high pH. Association constants and extinction coefficients have been evaluated at about pH 2 using the concentration dependence of this new band. The order of these constants is tryptamine > tryptophan > serotonin > indole (Table 5.6), which is different from the order of strength of interaction of the indoles with FMN. Freezing the indole–NAD complex gives rise to an increase in depth of the color. The spectral data clearly indicate a 1:1 stoichiometry, at least in solution. By analogy with the interactions with other molecules already discussed in this chapter, it is presumed that these complexes are charge-transfer complexes. It is recognized that certain other factors must contribute to the complex stability, since the order of complexing is not as predicted on the grounds of charge transfer alone. The site of electron acceptance of the coenzyme is thought to be primarily on the nicotinamide moiety of NAD (Fig. 5.9).

Cilento and his co-workers[47] have also examined the interaction of some models of indoles with NAD, and confirmed the spectral changes reported by Alivisatos *et al.* Measurements have been made of the associa-

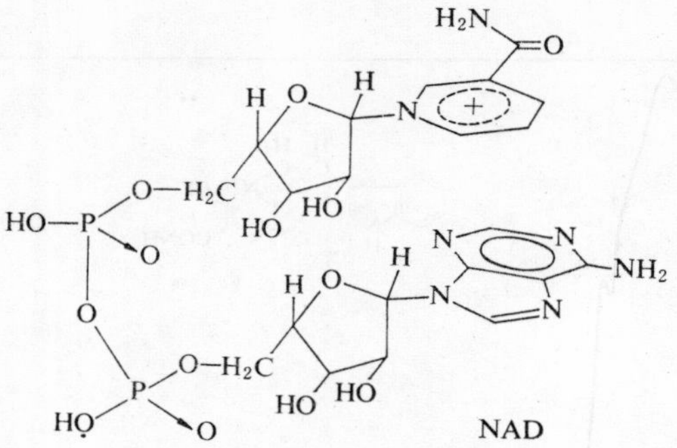

NAD

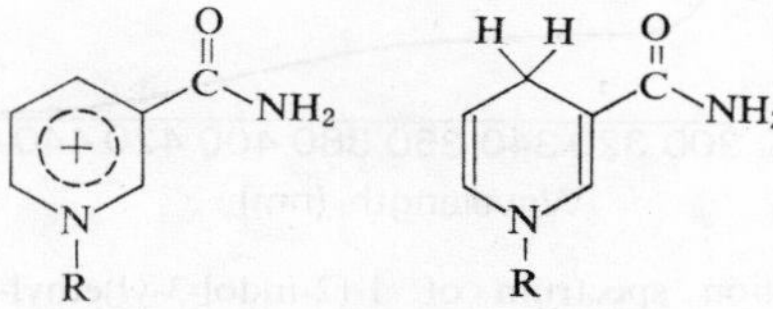

NAD NADH

Fig. 5.9 Nicotinamide-adenine dinucleotides.

TABLE 5.6 Data for indole–NAD complexes at 27°C

Compound	K_c (l/mole)	ε_{380} (cm^2/mole)	pH range
Serotonin†	11.10	367	2.0–2.4
Tryptamine	14.55	156	1.9–2.2
Tryptophan	13.40	151	1.9–2.5
Indole	4.13	430	1.9–2.4

† Utilized as the creatinine sulfate complex.

tion constants of certain indoles with 1-benzyl-3-carboxyamide pyridinium chloride, which is used as a model for oxidized pyridine coenzymes. The association constants are in a different order from those quoted previously.

A further, interesting suggestion has also been made by Cilento *et al.* They proposed that complex formation with tryptophan might be applicable for detecting exposed tryptophyl residues in proteins.

A series of model compounds in which indoles have been directly incorporated into pyridinium chlorides has been synthesized and studied by Shifrin.[48)] The spectrum of indolylethylnicotinamide in methanol is different from the spectrum of a 1:1 mixture of tryptamine hydrochloride and nicotinamide methochloride. There is an increase in absorption in the re-

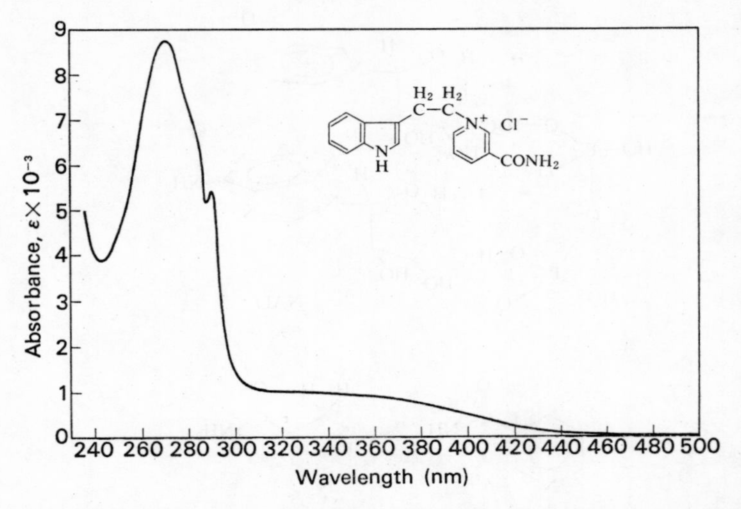

Fig. 5.10 Absorption spectrum of 1-(2-indol-3-yl)ethyl-3-carbamoylpyridinium chloride in methanol.
(Source: S. Shifrin, *Biochim. Biophys. Acta*, **81**, 205 (1964). Reproduced by kind permission of ASP Biological and Medical Press, Netherlands.)

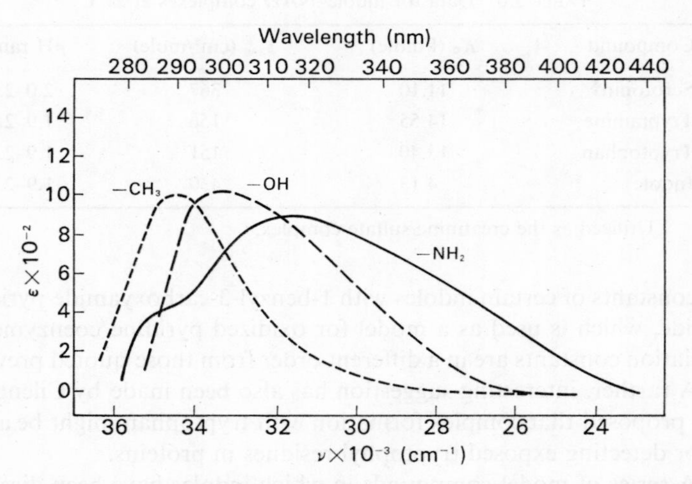

Fig. 5.11 Intermolecular charge-transfer spectra of *para*-substituted phenylethylnicotinamides in which the substituent is amino (—), hydroxyl (---) or methyl (....), respectively.
(Source: S. Shifrin, *Biochim. Biophys. Acta*, **96**, 173 (1965). Reproduced by kind permission of ASP Biological and Medical Press, Netherlands.)

gion of 320 nm in the model compound, thought to be a charge-transfer band (*cf.* Fig. 5.10)

Fluorescence studies of these compounds have produced interesting results. When indolylethyldihydronicotinamide is excited in the region of the indole absorption, the fluorescence spectrum of the molecule is that of indolylethyldihydronicotinamide, indicating that the excitation energy is very efficiently transferred from the indole moiety. Indolylethylnicotinamide, on the other hand, shows no emission from the indole moiety. The new absorption in the model compound is attributed to intramolecular charge transfer since the indoles are known charge donors and the pyridinium ring has good acceptor properties.

Further confirmation of the nature of the interaction in indolylethylnicotinamide has also been obtained by Shifrin,[49] who studied the behavior of compounds containing both *para*-substituted phenyl moieties and the 3-carbamidopyridinium system incorporated in the same molecule (VIII).

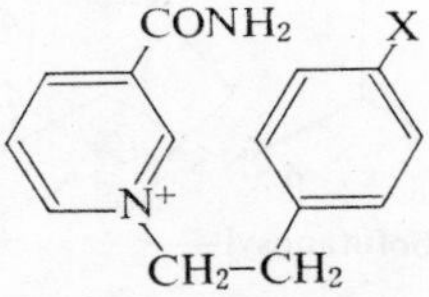

$CONH_2$ X

N^+

$CH_2{-}CH_2$

(VIII)

The absorption spectra of methanolic solutions of VIII, where X is NH_2, OH, OCH_3, CH_3, Cl or H, were compared with the absorptions of equimolar mixtures of the corresponding 4-X-phenylethylamines and 1-methylnicotinamide perchlorate (1-methyl-3-carbamidopyridinium perchlorate) in methanol (Fig. 5.11). The difference spectra have maxima, the energies of which are proportional to the ionization potentials of the corresponding substituted benzenes, C_6H_5X. This correlation, together with the observation that the intensities of these transitions are concentration-independent, suggests that the absorptions are the result of charge-transfer interactions across space between the pyridinium ring and benzene moiety in each of these compounds (see also Chapter 2).

The intramolecular interaction between the flavin and adenine moieties in FAD has been well-established by the work of Weber.[50] Various studies have been carried out to establish the structure of this molecule (Fig. 5.12). The general consensus appears to be that the two rings are stacked in a paralled mode.[51] However, a study using circular dichroism techniques has suggested that the two moieties are in close proximity but in a nonplanar arrangement. The agreement of experimentally determined

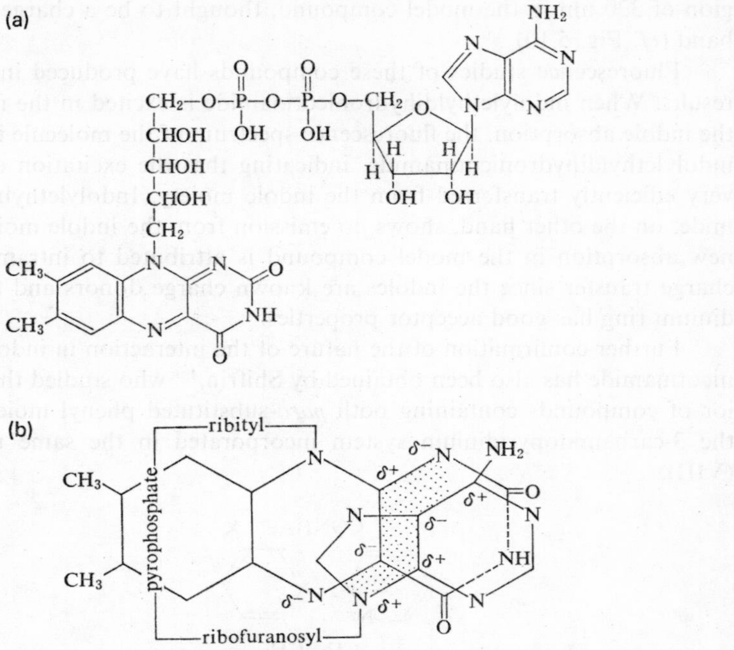

Fig. 5.12 The suggested structure for FAD (a), showing the possible structure for a stacked conformation of FAD (b). The essential feature of the indicated stacking is that the adenine moiety underlies the isoalloxazine moiety.

spectral energies and rotational strengths with theoretical values computed by Song[52] using the Miles and Urry[53] model is excellent.

Cilento *et al.* have also reported[54] that the broad absorption shown by solutions of glyceraldehyde-3-phosphate dehydrogenase containing NAD^+ may be the result of charge-transfer interaction, again between the pyridinium ring acting as the acceptor site and an indole moiety acting as the donor site.

Evidence for complex formation between pyridinium ions and neutral aromatic hydrocarbons has also been given recently.[55] The complexes appear to be charge-transfer complexes. In those formed between porphyrins and a variety of 1-substituted nicotinamides in aqueous solution, it is claimed that the effect of charge transfer is small. This contrasts with the interaction of porphyrins with certain biologically inactive compounds.

Flavoproteins

Strittmatter[56a] has investigated the nature of flavins binding in the flavoprotein, cytochrome b_5 reductase. FMN binding to this enzyme is severely inhibited in the presence of strong electron acceptors such as trinitrobenzsulphonate and iodine. The reductase also forms complexes with NADH, and less strongly with NAD^+. Association constants of the complexes with NADH, NADPH and their derivatives have been determined from fluorescence quenching and competitive binding studies. These association constants are orders of magnitude higher than those for the free flavins.

The interaction of the flavoprotein, lipoyl dehydrogenase, with the coenzymes NADH, TNADH, NHDH and NADPH has been described by Massey and Palmer.[56b] The spectrum of the enzyme exhibits long-wavelength bands which are associated with conditions conducive to full reduction of the enzyme (Fig. 5.13). These bands are dependent on the presence of NAD^+, since the effect of adding oxidized coenzymes on the spectrum of the reduced coenzyme is to cause the reappearance of the long-wavelength bands, which gives this enzyme its characteristic green color. These new long-wavelength bands are designated as charge-transfer bands with reduced enzyme as the donor.

Burton[57] has shown that the binding of FAD to the enzyme, D-amino

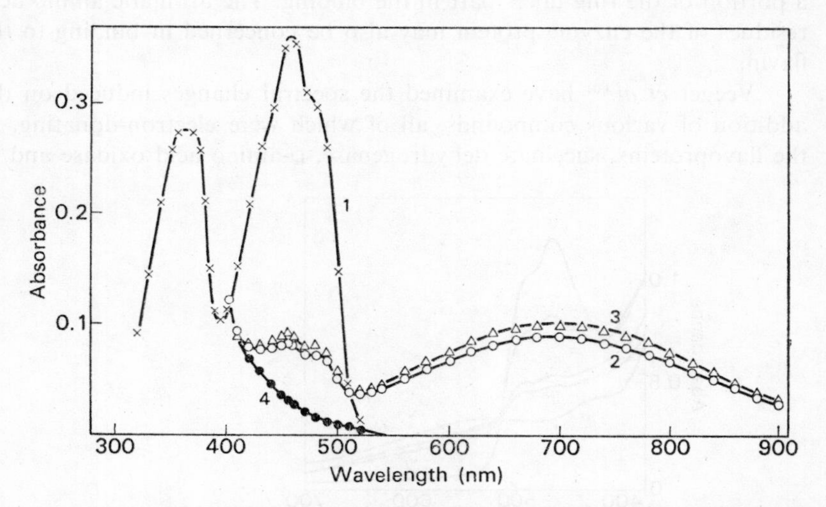

Fig. 5.13 Reduction of flavoprotein on addition of NADH to the oxidized enzyne. *Curve 1*: 3.2×10^{-5} M oxidized enzyme in 0.06 M phosphate, pH 6.3, plus 10^{-3} M sodium arsenite (25°C). *Curve 2*: plus 3.3×10^{-4} M NADH. *Curve 3*: after the further addition of 1.67×10^{-4} M FAD^+. *Curve 4*: after the further addition of NADase.
(Source: V. Massey and G. Palmer, *J. Biol. Chem.*, **237**, 2347 (1962). Reproduced by kind permission of the American Society of Biological Chemists, Inc., U.S.A.)

acid oxidase, is competitive with various purines. Certain of the competing purines (such as caffeine) inhibit the enzyme but purines related to FAD (such as AMP) though strongly competing with FAD in fact protect the enzyme. The inhibition of the enzyme by some antagonists in the presence of FAD is thought to be due to the formation of nonfluorescent complexes between the inhibitors and FAD.

Yagi and Ozawa[58] have examined the competitive binding of flavins and purines to the enzyme, apo-D-amino acid oxidase, and found that both flavin and purine compete with FAD but not with each other. This suggests that both the flavin and adenosine moieties of FAD bind to the enzyme. Yagi *et al.*[59] have also found that *p*-aminosalicylic acid inhibits the activity of the enzyme by forming a complex with FAD as well as by competitive inhibition with FAD and with the substrate. Similar results have been obtained for a large number of phenols.

Studies on the binding of flavin phosphates to some FMN-dependent enzymes have been carried out by Tsibris, McCormick and Wright.[51] From inhibition or reactivation studies of the enzymes or flavins, they showed that there is no specific dependence on the presence of groups at special sites in the alloxazine rings, but suggested that the whole or at least a portion of the ring takes part in the binding. The aromatic amino acid residues of the enzyme protein may also be concerned in binding to the flavin.

Veeger *et al.*[60] have examined the spectral changes induced on the addition of various compounds, all of which were electron-donating, to the flavoproteins, succinate dehydrogenase, D-amino acid oxidase and L-

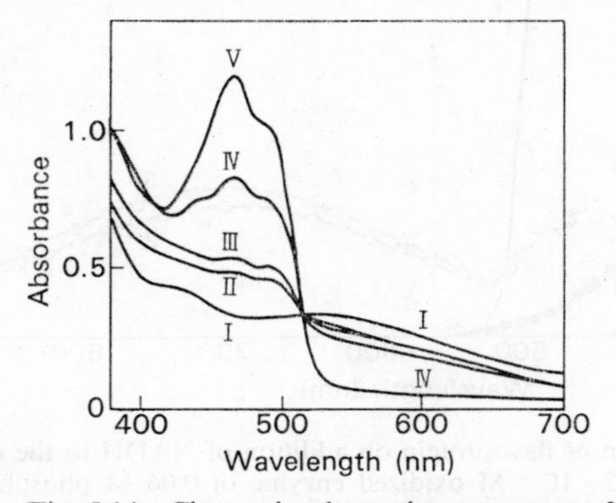

Fig. 5.14 Change in absorption spectrum of the purple intermediate upon mixing with known concentrations of sodium benzoate. I: the purple intermediate (1.06×10^{-4} M with respect to FAD). II–V: amount of sodium benzoate added to I (final concentration), 5.0×10^{-3} M (II), 1.0×10^{-2} M (III), 5.0×10^{-2} M (IV), or 1.0×10^{-1} M (V). All operations were carried out under anaerobic conditions.

amino acid oxidase. They stated that the changes arise from the formation of charge-transfer complexes.

Yagi *et al.*[61] have investigated the interaction of D-amino acid oxidase, and its purple complex with amino acids, with sodium benzoate. The spectrum of the enzyme–benzoate complex was the same as the spectrum arising from the interaction of the purple complex with benzoate (Fig. 5.14). "Old" purple complex, which is believed to change to the semiquinoid form of the enzyme, shows only a partial change from the spectrum of the enzyme–benzoate complex and there is a blue coloration. This is ascribed to the partial formation of a semiquinoid-enzyme–benzoate complex which reverts fully to the enzyme–benzoate complex on the introduction of oxygen.

5.3 Electron Donors and Acceptors of Phamaceutical Interest

Quinones

A series of experiments has been carried out on the complexes formed between menadione and various biochemical electron donors of pharmaceutical interest by Hata and his colleagues.[62]

Very few charge-transfer studies appear to have been made on biologically active quinones. In general, it is anticipated that quinones will act as electron acceptors, although this property in a particular quinonoid system will depend very much upon the nature and extent of substitution.

The compound 2-methyl-1,4-naphthoquinone, also called vitamin K_3,

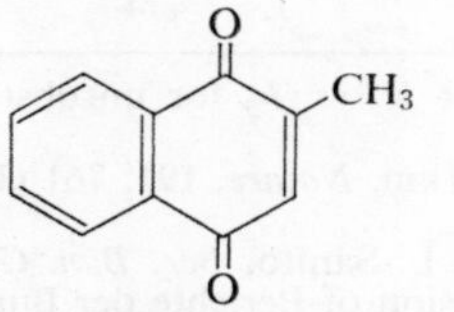

O
CH3
O

vitamin K_3 (menadione)

which was at one time used under the name menadione as a synthetic substitute for vitamin K_1, has been shown by Cilento and Sanioto[63] to complex with aromatic hydrocarbons. The general behavior of these systems suggests that charge-transfer complexation is involved, although exact details are unclear because of the low stability of the complexes under the conditions used in these studies. Recent determinations of the stabilities of complexes of 2-methyl-1,4-naphthoquinone with a porphyrin and with chlorophyll by Williams and his co-workers[64)] have been made by the spectroscopic method. Complex formation between the antioxidant, *N*-phenyl-2-naphthylamine, and vitamin K_3 has also been described recently.[65)]

Cilento *et al.* have reported[66)] that the addition of various aromatic hydrocarbons as donors to menadione in water causes a decrease in absorption of the acceptor band but an increase in absorption on the long-wavelength side. Association constants and thermodynamic parameters have been evaluated for the complexes, as have the energies of the highest filled molecular orbitals of the donors (Table 5.7).

TABLE 5.7 Data for complexes of polycyclic aromatic hydrocarbons with menadione in solution in $CHCl_3$ at 25°C ± 2°C†[1]

Hydrocarbon	λ (nm)	K_c (M^{-1})	I_p(eV)†[2]
7,12-Dimethyl-1,2-benzanthracene	478	0.36	—
7-Methyl-1,2-benzanthracene	476	0.40	7.37
10-Methyl-1,2-benzanthracene	478	0.42	7.37
20-Methylcholanthrene	500	0.43	
4-Methyl-1,2-benzanthracene	474	0.45	7.41
1,2,5,6-Dibenzanthracene	472	0.50	7.80
Anthracene	472	0.55	7.37
2′-Methyl-1,2-benzanthracene	476	0.60	7.39
3,4-Benzpyrene	510	0.69	7.19
Tetracene	484	5.27†[3]	7.00
3-Methyl-1,2-benzanthracene	476	0.0	7.43
6-Methyl-1,2-benzanthracene	472	0.4	7.37
3′-Methyl-1,2-benzanthracene	472	0.4	7.43
1,2-Benzanthracene	474	0.6	7.45

†[1] Correlation coefficient for K_c *vs.* I_p for unsubstituted hydrocarbons = −0.87 at 96% confidence level.
†[2] J. B. Birks and M. A. Slifkin, *Nature*, **191**, 761 (1961).
†[3] At 10°C.
(Source: G. Cilento and D. L. Sanito, *Ber. Bun. Ges. Phys. Chem.*, **67**, 426 (1963). Reproduced by kind permission of Berichte der Bunsengesellschaft für Physikalische Chemie, W. Germany.)

Menadione readily decomposes on exposure to near-ultraviolet light. The addition of electron donors to aqueous menadione solutions helps to stabilize the acceptor.[67)] The degree of stabilization is proportional to both the free energy of the complexes and the enthalpy of dissociation. It is thus suggested that this stabilization of menadione against photodecomposition is due to the formation of charge-transfer complexes. Molecular orbital calculations on the complexes have given energies of stabilization which correlate linearly with the experimental enthalpies of dissociation.[68] The charge-transfer maxima of the complexes in aqueous solution correlate linearly with thermodynamic parameters and the stabilization against photodecomposition.[69]

PARAQUAT (METHYLVIOLOGEN)

Paraquat (methylviologen) has interesting biochemical properties. It inhibits electron transfer both in the cytochrome chain of mitochondria and in the electron transfer chain of chloroplasts. It is in widespread use as a weed killer.

$H_3C\text{-}N^{+}$ ... $^{+}N\text{-}CH_3\ 2Cl^{-}$

paraquat

The absorption spectra of various paraquat salts show bands which are assigned to intermolecular transfer between the anion and the cation.[70] Studies on these salts by esr spectrometry have led to the conclusion that complete electron transfer occurs in at least some of the salts. Additional evidence for this is the high electrical conductivity shown by the salts.

Complexes of paraquat dichloride and various π- and n-electron donors have been investigated by various methods (Table 5.8). The paraquat salt forms 1:1 complexes with the donors in solution. Charge-transfer bands are observed and association constants have been evaluated using the Benesi-Hildebrand equation. The association constants of hydroquinone–paraquat dichloride complexes have also been evaluated by nmr methods and the results are identical with those obtained spectrophotometrically. The enthalpy of dissociation of the hydroquinone complex is 5.45 kcal/mole. In the solid state, isolated complexes have a 2:1 stoichiometry. This difference in stoichiometry between the solution and solid state is not too unusual: for example, similar differences have been noted in phenothiazine complexes. In solution, the formation of a 1:1 complex lowers the electron positivity of the paraquat, rendering it a poor electron acceptor. In the solid state, the crystalline forces help to stabilize the complex in a 1:2 mode. Prout *et al.*[71] have made a complete crystalline structure analysis of the paraquat–$CoCl_4$ charge-transfer complex (Fig. 5.15).

Positive identification of the charge-transfer bands has been made

TABLE 5.8 Data for complexes of paraquat (methylviologen)

Donor	$K_c(M^{-1})$	λ_{max}	T (°C)
N-Phenyl-2-naphthylamine	88	550	20
Ferrocyanide	52	~600	23
Methylviologen ferrocyanide	0.21		23
Hydroquinone ($\Delta H = -5.45$ kcal/mole)	5.23	440	25
Pyrene	3.4	465	25
p-Aminophenol	7.7	495	25
o-Phenylenediamine	14.1	525	25
α-Naphthylamine	8.7	525	25
p-Phenylenediamine	15.4	555	25
Chlorpromazine ($\Delta H = -3$ kcal/mole)	15	500	28
3-Indoleacetate ($\Delta H = -2.5$ kcal/mole)	6	390	28
3-Indolebutyrate ($\Delta H = -2.5$ kcal/mole)	13.5	400	28
α-Naphthoate	9.1	375	28
α-Naphthol	13.2	433	28

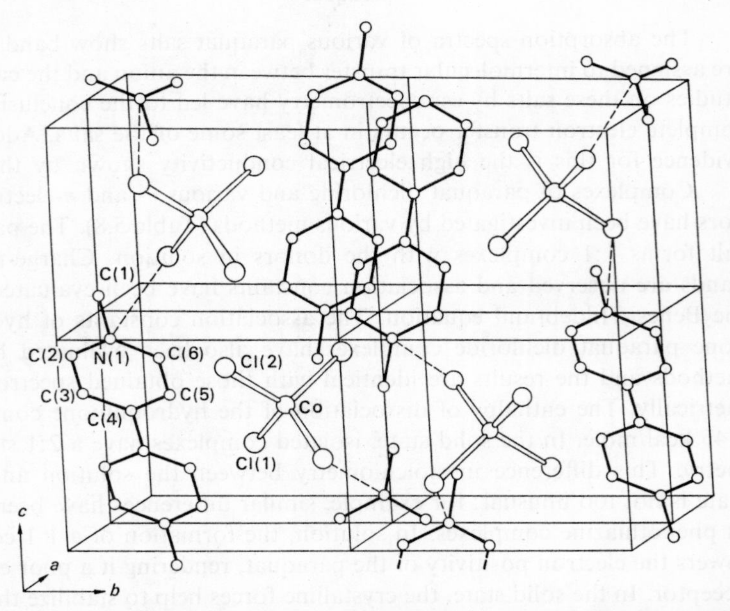

Fig. 5.15 Clinographic projection of the paraquat(pq)–$CoCl_4$ CT complex crystal structure.
(Source: C. K. Prout and P. Murray-Rust, *J. Chem. Soc.* (A), **1969**, 1520. Reproduced by kind permission of the Chemical Society, England.)

TABLE 5.9 A short compilation of acceptor electron affinity evaluations, which are thought to be comparable

Electron acceptor	Electron affinity (eV)	Ref.
1,2,4,5-Tetracyanobenzene	0.4	(a)
Maleic anhydride	0.4	(a)
S-Trinitrobenzene	0.6	(b)
p-Benzoquinone	0.8	(b)
Pyromellitic dianhydride	0.85	(a)
Indanetrione	1.10	(c)
Paraquat	1.24	
Chloranil	1.35	(b)
Tetracyanoethylene	1.8	(a)

(a) G. Briegleb and J. Czekalla, *Angew. Chem.*, **72**, 40 (1960).
(b) G. Briegleb, *ibid.*, **76**, 326 (1964).
(c) A. R. Lepley and J. S. Thelman, *Tetrahedron*, **22**, 101 (1966).

from the linear relationship between the position of the bands in the paraquat complexes and in the charge-transfer complexes of the same donors with *p*-benzoquinone, chloranil, trinitrobenzene and tetracyanoethylene.[72] The electron affinity of paraquat computed by correlating the charge-transfer band positions with the energies of the calculated highest occupied molecular orbitals of the donors is 1.24 eV (Table 5.9).

DRUG INTERACTIONS

Little progress has yet been made on the general problem of drug action at the molecular level. Some substances such as anesthetics and other lipotropic drugs often show a low specificity and may well act by an essentially "physical" mechanism. However, such compounds are atypical in that the majority of physiologically active compounds foreign to the biochemical substrate show a high degree of specificity. This particular characteristic is even more prominent in the case of natural chemical substances.

The basic process of stimulation at the molecular level has given rise to modifications of the "receptor". Some workers consider that a drug acts so long as it is attached to the "receptor" but that different drugs have different efficacies when they occupy the receptor site. An alternative explanation of the differences shown by two drugs at the same receptor is that the activation of the receptor depends upon the rate of encounter with a drug at the receptor. Such problems have caused pharmacologists to turn their attention to the relationship between the chemical structure of drugs and their physiological action. These studies, which can generally be made at least semi-quantitative, may lead to a better understanding of the molecular nature of drug-receptor interactions.

In many cases, it would appear that the binding of the drug to the receptor is of relatively low energy, certainly considerably less than that involved in normal covalent bonding. Chemically, "receptor-drug complexing" could occur by ionic association, through hydrogen bonding, by other weaker forces which might include charge-transfer forces, or, as is more likely, a combination of several of these types of forces. As a first step to determine whether charge-transfer forces are in any way involved, the ability of drugs and related compounds to form charge-transfer complexes with well-characterized electron acceptors or electron donors, mainly in non-aqueous solvents, has been studied. Evidence for complexing by indoles with electron acceptors has already been mentioned.

Considerable interest has been shown in the phenothiazine drugs. Various workers[73] have noted the low ionization potentials of these compounds. Calculations of the highest occupied molecular orbital of phenothiazine (IX) and of chlorpromazine (X) also indicate that these molecules should be strong electron donors.[74]

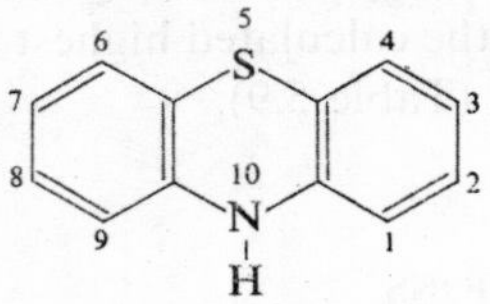

(IX) 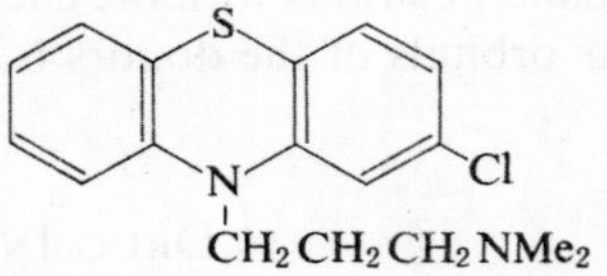 (X)

When the energies of the maxima of such bands are plotted against energy maxima for the corresponding hexamethylbenzene-electron-acceptor complexes, straight lines are obtained (Fig. 5.16) which are typical of charge-transfer complexes (see Chapter 2). The equilibria have been measured for various phenothiazine drugs, including promethazine (XI) ethopropazine (XII) diethazine (XIII), trimeprazine (XIV) and promazine (XV), with the electron acceptor, 1,4-dinitrobenzene in chloroform and in carbon tetrachloride solution.[75]

The results show that, relative to *N,N*-dimethylaniline, these compounds are good electron donors and form 1:1 complexes with electron acceptors. Biochemically inactive 10-methylphenothiazine has a higher association constant than any of the drug molecules measured, all of which possess a large side-chain in the 10-position. This suggests that complex formation may be reduced by steric interference from this side-chain in

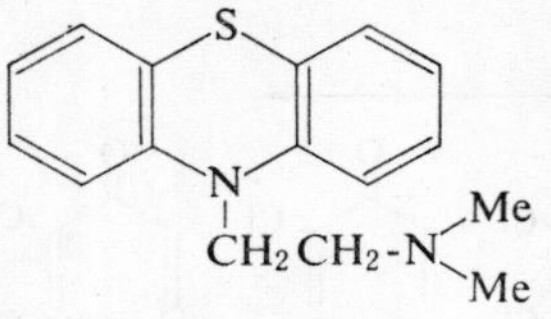

(XI) 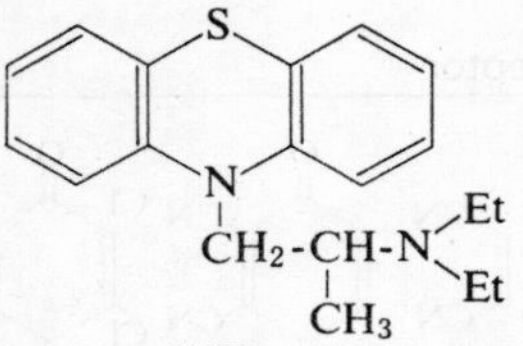 (XII)

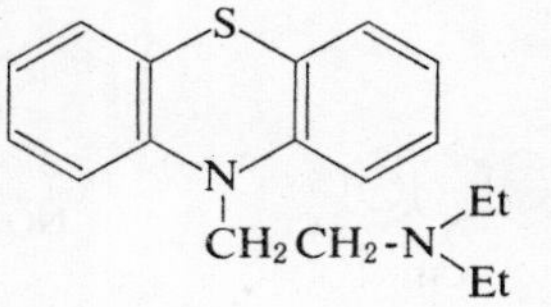

(XIII) 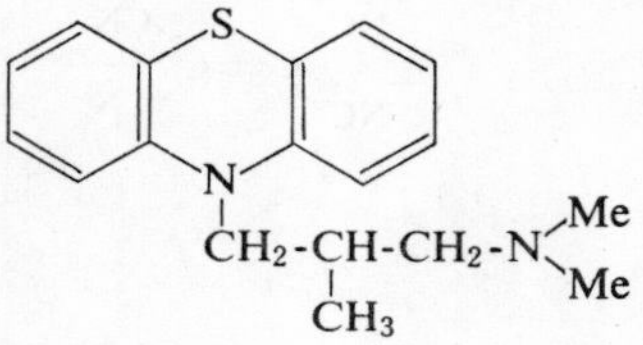 (XIV)

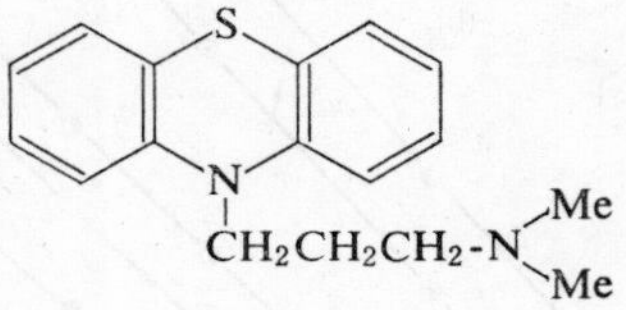

(XV)

these particular complexes. By contrast, a double role for phenothiazine drug molecules at a biological site of action has been suggested, whereby the drug acts as a π-electron donor in a charge-transfer interaction with one receptor and as a hydrogen-bonding agent, via a protonated aliphatic amino nitrogen atom in the side-chain, at a second receptor. These two suggestions are not necessarily conflicting. They do illustrate, however, the care needed in drawing conclusions about possible biochemical behavior based on observations of model systems, particularly those not intended to represent the drug-receptor complex.

The spectrum of a mixture of FAD and chlorpromazine shows a complete loss of flavin absorption in the region of 450–500 nm.[76] The fluorescence of the flavin in the mixture is very strongly quenched. The association constant derived from quenching studies using the Stern-Volmer equation is *ca.* 1000 l/mole. Very similar results have also been obtained for complexes of RFN with the drug. The view that chlorpromazine forms a strong complex with FMN is supported by Kerreman, Isenberg and Szent-Györgyi[74] who found at $-70°C$ a new absorption band at 490 nm, which they believe to be the band of the semiquinone stabilized by

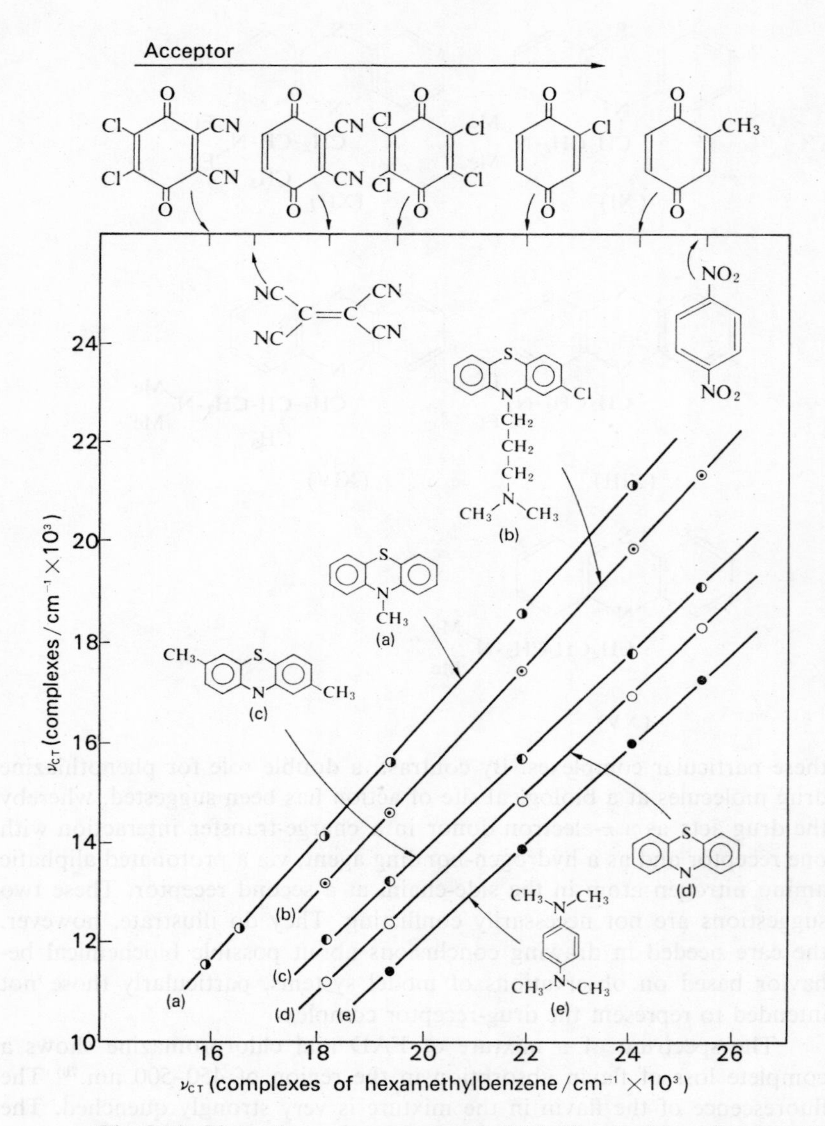

Fig. 5.16 Plots of ν_{CT} for complexes of (a) 10-methylphenothiazine, (b) chlorpromazine, (c) 3,7-dimethylplenothiazine, (d) phenothiazine, and (e) *N,N,N′,N′*-tetramethyl-*p*-phenylenediamine with various acceptors; all against ν_{CT} for the corresponding hexamethylbenzenes.
(Source: R. Foster, *Organic Charge-Transfer Complexes*, p. 359, 1969. Reproduced by kind permission of Academic Press Inc. (London) Ltd., England.)

charge-transfer, as well as a broad band at 570 nm similar to that exhibited by indole–flavin complexes.

Cann[77] has demonstrated that chlorpromazine forms 1:1 complexes with hemin, hematoporphyrin, and myoglobin. A similar complex is also formed with paraquat (methylviologen).

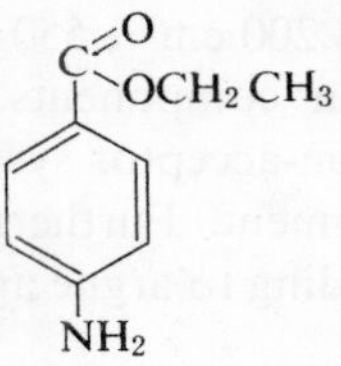

benzocaine 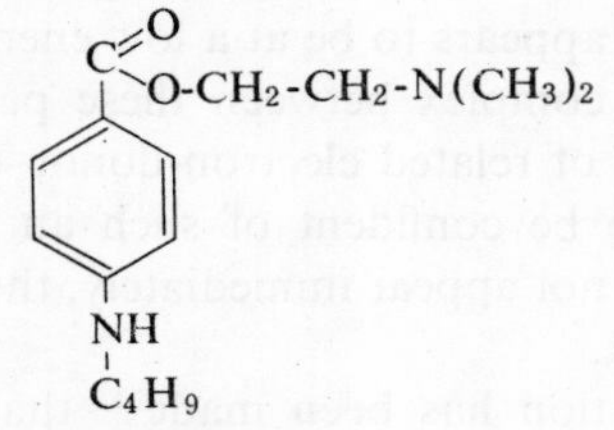 retracaine

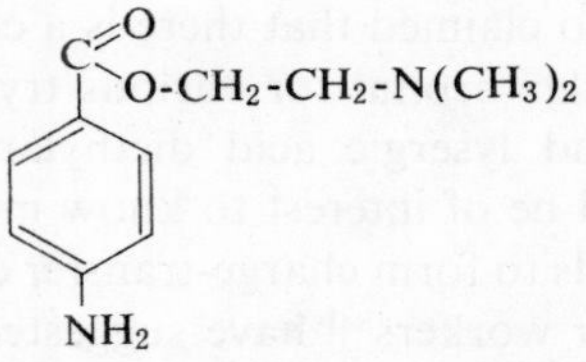

procaine 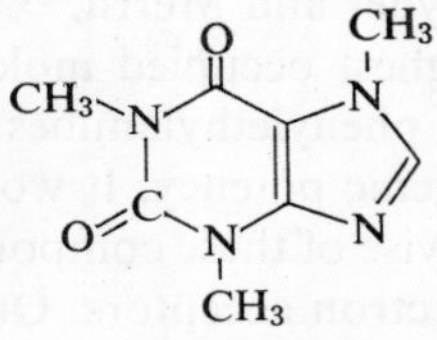 caffeine

Higuchi and Lachman[78] have studied the effect of caffeine on the alkaline hydrolysis of *p*-aminobenzoic ethylate (benzocaine) at 30°C, $[Ba(OH)_2]=0.04$ N. It was found that 2.5% caffeine retards the rate of hydrolysis 1/5.4 times. Benzocaine complexed with caffeine is deactivated to the attack of OH^-. The stability constant of the complex between benzocaine and caffeine has been measured as 60 l/mole, in good accord with that calculated from the kinetic data. Lachman, Ravin and Higuchi[79] also examined the effect of caffeine on the hydrolysis of procaine (*p*-aminobenzoic-*N*,*N*-dimethylaminoethylate). The 2:1 complex between caffeine and procaine has a stability constant of 620 l/mole in water and 30% benzene, 70% *iso*-octane (v/v%). The addition of 50% caffeine causes a decrease in the hydrolysis rate of 2.7 times compared to free procaine. To a first approximation, the complexed procaine does not react, so that this method would be effective for the preservation of its medical effectiveness by complexation. With tetracaine (*p*-(*N*-butylamino)benzoic-2-*N*,*N*-diaminoethylate), caffeine showed effective retardation of the hydrolysis.[80] For example, 2.5% caffeine reduced the rate of hydrolysis of tetracaine about 3.5 times compared with the free molecule; the constant of 2:1 complex formation between caffeine and tetracaine is 900 $l^2/mole^2$.

Another development is the study of interactions which are anticipated to be more akin to the drug-receptor complex. A knowledge of the behavior of the interaction of drugs with amides, including simple polypeptides, would be of value. Agin[81)] has reported that a transient magenta color develops when procaine hydrochloride is mixed with RNA. He suggests that this is the result of charge-transfer complex formation. The band, however, appears to be at a low energy (18,200 cm^{-1}, 550 nm) for a charge-transfer complex between these particular components. Without the comparison of related electron-donor–electron-acceptor systems, it is thus difficult to be confident of such an assignment. Furthermore, the coloration does not appear immediately, thus tending to argue against this type of complex.

The suggestion has been made[74)] that the mechanism of action of chlorpromazine and other psychotropic drugs may in some way be due to the electron-donating ability of these compounds. This idea has been extended by Snyder and Merril,[82)] who claimed that there is a correlation between the highest occupied molecular orbital for various tryptamines, amphetamines, phenylethylamines and lysergic acid diethylamide, and their hallucinogenic potency. It would be of interest to know more of the ability or otherwise of these compounds to form charge-transfer complexes with various electron acceptors. Other workers[83)] have suggested that the psychotropic activity of chlorpromazine is due in fact to the positive ion radical formed by the loss of a single electron. Evidence in support of its intercalation in the DNA helix has been given.[84)]

REFERENCES

1. M. Calvin, *Rev. Mod. Phys.*, **31**, 147 (1949); W. Arnold and R. K. Clayton, *Proc. Natl. Acad. Sci., U.S.A.*, **46**, 769 (1960).
2. D. R. Kearns and M. Calvin, *J. Am. Chem. Soc.*, **83**, 2110 (1961).
3. M. Calvin, *J. Theoret. Biol.*, **1**, 258 (1961).
4. J. R. Platt, *Science*, **129**, 372 (1959).
5. D. R. Kearns, G. Tollin and M. Calvin, *J. Chem. Phys.*, **32**, 1020 (1960).
6. G. Tollin and G. Green, *Biochim. Biophys. Acta*, **60**, 524 (1962).
7. G. Tollin and G. Green, *ibid.*, **66**, 308 (1963).
8. W. R. Boon, *Endeavour, XXVI*, **97**, 27 (1967).
9. R. E. Kofahl and H. G. Lucas, *J. Am. Chem. Soc.*, **76**, 3931 (1954).
10. A. Pullman, *Ann. Chim* (*Paris*), **2**, 5 (1947).
11. K. H. Takemura, M. D. Cameron and M. S. Newman, *J. Am. Chem. Soc.*, **75**, 3280 (1953).
12. A. Pullman and B. Pullman, *Cancérisation par les Substances Chimiques et Structure Moleculaire,* Masson, 1955.

13. S. S. Epstein, I. Bulon, J. Koplan, M. Small and N. Mantel, *Nature*, **204**, 750 (1964).
14. J. E. Love Lock, A. Zlatkis and R. S. Beckev, *ibid.*, **193**, 540 (1962); A. C. Allison and J. W. Lightbroun, *ibid.*, **189**, 892 (1961).
15. A. Pullman and B. Pullman, *ibid.*, **196**, 228 (1962).
16. A. C. Allison and T. Nash, *ibid.*, **199**, 469 (1963).
17. B. Pullman and A. Pullman, *ibid.*, **199**, 467 (1963).
18. J. B. Birks, *ibid.*, **190**, 232 (1961).
19. A. Pullman, *Biopolymers Symp.*, **1**, 47 (1964).
20. C. Heidelberger and M. G. Moldenhauer, *Cancer Res.*, **16**, 442 (1956).
21. V. T. Oliverio and C. Heidelberger, *ibid.*, **18**, 1094 (1958).
22. P. M. Bhargava and C. Heidelberger, *J. Am. Chem. Soc.*, **77**, 2877 (1955); **78**, 3671 (1956).
23. A. Pullman and B. Pullman, *Advan. Cancer Res.*, **3**, 117 (1955).
24. C. A. Coulson, *ibid.*, **1**, 1 (1953).
25. G. M. Badger, *J. Chem. Soc.*, **1949**, (1949); **1950**, 1806.
26. J. Booth and E. Boyland, *Biochim. Biophys. Acta*, **12**, 75 (1953).
27. A. M. Liquori, B. DeLerma, B. Ascoi, C. Botre and M. Transciatti, *J. Mol. Biol.*, **8**, 20 (1964).
28. P. O. P. T'su and P. Lu, *Proc. Natl. Acad. Sci. U.S.A.*, **51**, 17 (1964).
29. E. Boyland and B. Green, *Brit. J. Cancer*, **16**, 507 (1964).
30. R. S. Snart, *Trans. Faraday Soc.*, **63**, 2384 (1967).
31. N. Brock, H. Druckrey and H. Hamperl, *Arch. Exp. Path. Pharm.*, **189**, 709 (1938).
32. H. Weil-Malherbe, *Biochem. J.*, **40**, 351 (1946).
33. A. Damiani, E. Giglio, A. M. Liquori and A. Ripamonti, *J. Mol. Biol.*, **20**, 211 (1966); **23**, 113 (1967).
34. J. Booth, E. Boyland, D. Manson and G. H. Wiltshire, *Rept. Brit. Emp. Cancer Campaign*, **29**, 27 (1931).
35. L. C. Kanner and L. M. Kozloff, *Biochemistry*, **3**, 215 (1964).
36. E. M. Kosower, *Progr. Phys. Org. Chem.*, **3**, 143 (1965).
37. H. Beinert, *J. Am. Chem. Soc.*, **47**, 114 (1956).
38. E. M. Kosower, *Flavins and Flavoproteins*, Elsevier, 1965.
39. K. Yagi, A. Kotati, M. Naoi and K. Okamura, *J. Biochem*, **60**, 236 (1966); K. Yagi, K. Okamura, M. Naoi, A. Takai and A. Kotaki, *ibid.*, **66**, 581 (1969).
40. M. A. Slifkin and A. C. Allison, *Nature*, **215**, 949 (1967).
41. I. Isenberg and A. Szent-Györgyi, *Proc. Natl. Acad. Sci. U.S.A.*, **44**, 857 (1958).
42. H. A. Harbury and K. A. Foley, *ibid.*, **44**, 662 (1958).
43. G. K. Radda and M. Calvin, *Biochemistry*, **3**, 384 (1964).
44. I. Isenberg, S. L. Baird and A. Szent-Györgyi, *Proc. Natl. Acad. Sci. U.S.A.*, **47**, 245 (1961).
45. A. I. Popov, C. C. Bisi and M. Craft, *J. Am. Chem. Soc.*, **80**, 6513 (1958).
46. S. G. A. Alivisatos, F. Unger, A. Jibril and G. A. Mourkides, *Biochim. Biophys Acta*, **51**, 361 (1961).
47. G. Cilento and P. Guisti, *J. Am. Chem. Soc.*, **81**, 3801 (1959); G. Cilento and P. Tedeschi, *J. Biol. Chem.*, **236**, 907 (1964).
48. S. Shifrin, *Biochim. Biophys. Acta*, **81**, 205 (1963).
49. S. Shifrin, *ibid*, **96**, 173 (1965)
50. G. Weber, *Biochem. J.*, **47**, 114 (1950).
51. J. C. M. Tsibris, D. B. McCormick and L. D. Wright, *Biochemistry*, **4**, 504 (1965).
52. P-S. Song, *J. Am. Chem. Soc.*, **91**, 1850 (1969).
53. D. W. Miles and D. W. Urry, *Biochemistry*, **7**, 2791 (1968).
54. G. Cilento and P. Tedeschi, *J. Biol. Chem.*, **236**, 907 (1961).
55. G. Cilento and D. L. Sanioto, *Arch. Biochem. Biophys*, **110**, 133 (1965).
56.(a) P. Strittmatter, *J. Biol. Chem.*, **236**, 2329, 2336 (1961).
(b) V. Massey and G. Palmer, *ibid.*, **237**, 2347 (1962).
57. K. Burton, *Biochem. J.*, **48**, 458 (1951).

58. K. Yagi and T. Ozawa, *Biochim. Biophys. Acta*, **42**, 381 (1960).
59. K. Yagi and T. Ozawa, *ibid.*, **35**, 102 (1959).
60. C. Veeger, D. V. Der Vartian, J. F. Kale, A. de Kok and J. F. Koster, *Flavins and Flavoproteins*, p, 242, Elsevier, 1960.
61. K. Yagi, K. Okamura, M. Naoi, N. Sugiura and A. Kotaki, *Biochim. Biophys. Acta*, **146**, 77 (1967).
62. S. Hata, K. Mizuno and S. Tomioka, *Chem. Pharm. Bull.*, **15**, 1791, 1796 (1967); S. Hata and S. Tomioka, *ibid.*, **16**, 1397, 2078 (1968).
63. G. Cilento and D. L. Sanioto, *Z. Phys. Chem.*, **223**, 333 (1963); *Ber. Phys. Chem.*, **67**, 426 (1963).
64. H. A. O. Hill, A. J. Macfarlane, B. E. Mann and R. J. P. Williams, *Chem. Commun.*, **1968**, 123.
65. A. Ledwith and D. H. Iles, *Chem. Brit.*, **4**, 266 (1968).
66. G. Cilento and D. L. Sanioto, *Ber. Bun. Ges. Phys. Chem.*, **67**, 426 (1963).
67. S. Hata, K. Mizuno and S. Tomioka, *Chem. Pharm. Bull*, **15**, 1796 (1967).
68. S. Hata, *ibid.*, **16**, 1 (1968).
69. S. Hata and S. Tomioka, *ibid.*, **16**, 1397 (1968).
70. A. J. Macfarlane and R. J. P. Williams, *J. Chem. Soc.* (A), **1969**, 1517.
71. C. K. Prout and P. Murray-Rust, *ibid.*, **1969**, 1520.
72. B. G. White, *Trans. Faraday Soc.*, **65**, 2000 (1969).
73. D. R. Kearns and M. Calvin, *J. Phys. Chem. Ithaca*, **34**, 2026 (1961); L. E. Lyons and J. C. Mackie, *Nature*, **197**, 589 (1963).
74. G. Kerreman, I. Isenberg and A. Szent-Györgyi, *Science*, **130**, 1191 (1959).
75. R. Foster and P. Hanson, *Biochim. Biophys. Acta*, **112**, 482 (1966); R. Foster and C. A. Fyfe, *ibid.*, **112**, 490 (1966).
76. K. Yagi, T. Ozawa, and T. Nagatsu, *Nature*, **184**, 892 (1959).
77. G. R. Cann, *Biochemistry*, **6**, 3427 (1967).
78. T. Higuchi and L. Lachman, *J. Pharm. Sci.*, **44**, 521 (1955).
79. L. Lachman, L. J. Ravin and T. Higuchi, *ibid.*, **45**, 290 (1956).
80. L. Lachman and T. Higuchi, *ibid.*, **46**, 32 (1957).
81. D. Agin, *Nature*, **205**, 805 (1965).
82. S. H. Snyder and C. R. Merril, *Proc. Natl. Acad. Sci. U.S.A.*, **54**, 258 (1965).
83. L. H. Piette, G. Bulow and I. Yamazaki, *Biochim. Biophys. Acta*, **88**, 120 (1964).
84. S. Onishi and H. M. McConnell, *J. Am. Chem. Soc.*, **87**, 2293 (1965).

APPENDIX

P. Pfeiffer, *Die Organische Molekülarverbindungen*, Ferdinand Enke, 1927.

G. Briegleb, *Elektronen Donator-Acceptor Complexe*, Springer-Verlag, 1961.

R. S. Mulliken and W. B. Person, *Molecular Complexes*, Interscience, 1969.

A. Szent-Györgyi, *Introduction to a Submolecular Biology*, Academic Press, 1960.

L. J. Andrews and R. M. Keefer, *Molecular Complexes in Organic Chemistry*, Holden-Day Inc., 1964.

E. M. Kosower, *Molecular Biochemistry*, McGraw-Hill, 1962.

E. M. Kosower, *Progress in Physical Organic Chemistry*, vol. 3, p. 123, 1965.

T. Rose, *Molecular Complexes*, Pergamon Press, 1967.

R. Foster, *Organic Charge-Transfer Complexes*, Academic Press, 1969.

B. Pullman, *Electronic Aspects of Biochemistry*, Academic Press, 1964; The Structure and Properties of Biomolecules and Biological Systems, *Advances in Chemical Physics*, vol. VII, 1964.

M. Calvin, Quantum Conversion in Chloroplasts, *Advan. Catalysis*, **14**, 10 (1966); *Light and Life* (ed. W. D. McElroy and B. Glas), p. 317, John-Hopkins Press, 1961; *Rev. Mod. Phys.*, **31**, 147, 157 (1959).

M. A. Slifkin, *Charge-Transfer Interactions of Biomolecules*, Academic Press (London), 1971.

O. B. Nagy and J. B. Nagy, *Ind. Chim. Belg.*, **36**, 829, 929 (1971).

Author Index

Numbers in parentheses are reference numbers and indicate that an author's work is referred to although his name may not be cited in the text. Numbers in intalics show the page on which the complete reference is listed.

Subject Index

F, G

H

I

K, L

M

N

O

P

R

S

T

V, W, X

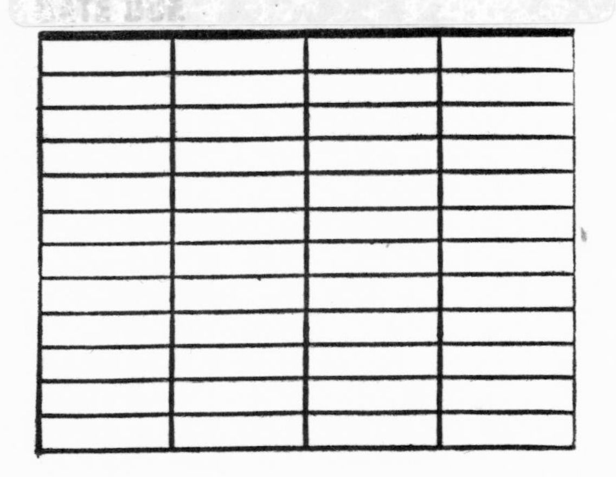

WHEATON COLLEGE LIBRARY
WHEATON, ILLINOIS